Introduction to Discrete-Time Signals and Systems

Introduction to Discrete-Time Signals and Systems

R.I. Damper

Department of Electronics and Computer Science
University of Southampton
UK

CHAPMAN & HALL

London · Glasgow · Weinheim · New York · Tokyo · Melbourne · Madras

003.83
D16i

Published by Chapman & Hall, 2–6 Boundary Row, London SE1 8HN, UK

Chapman & Hall, 2–6 Boundary Row, London SE1 8HN, UK

Blackie Academic & Professional, Wester Cleddens Road, Bishopbriggs, Glasgow G64 2NZ, UK

Chapman & Hall GmbH, Pappelallee 3, 69469 Weinheim, Germany

Chapman & Hall USA, 115 Fifth Avenue, New York, NY 10003, USA

Chapman & Hall Japan, ITP-Japan, Kyowa Building, 3F, 2-2-1 Hirakawacho, Chiyoda-ku, Tokyo 102, Japan

Chapman & Hall Australia, 102 Dodds Street, South Melbourne, Victoria 3205, Australia

Chapman & Hall India, R. Seshadri, 32 Second Main Road, CIT East, Madras 600 035, India

First edition 1995

© 1995 R.I. Damper

Printed in England by Clays Ltd., St Ives plc.

ISBN 0 412 47650 9

Apart from any fair dealing for the purposes of research or private study, or criticism or review, as permitted under the UK Copyright Designs and Patents Act, 1988, this publication may not be reproduced, stored, or transmitted, in any form or by any means, without the prior permission in writing of the publishers, or in the case of reprographic reproduction only in accordance with the terms of the licences issued by the Copyright Licensing Agency in the UK, or in accordance with the terms of licences issued by the appropriate Reproduction Rights Organization outside the UK. Enquiries concerning reproduction outside the terms stated here should be sent to the publishers at the London address printed on this page.

The publisher makes no representation, express or implied, with regard to the accuracy of the information contained in this book and cannot accept any legal responsibility or liability for any errors or omissions that may be made.

A catalogue record for this book is available from the British Library

Printed on permanent acid-free text paper, manufactured in accordance with ANSI/NISO Z39.48-1992 and ANSI/NISO Z39.48-1984 (Permanence of Paper).

In fond memory of my parents:

Wilfred Edward ('Ted') Damper
1916–1986

Ivy Damper
1917–1994

University Libraries
Carnegie Mellon University
Pittsburgh PA 15213-3890

Contents

Preface

This book is primarily intended to be used in conjunction with a first course in discrete-time signals and systems, or digital signal processing, within an electrical and/or electronic engineering degree scheme. It is expected that such a course will usually be taken at second year level within a three- or four-year scheme, although some of the later material can sensibly be left for third-year study. The reader is assumed to have a working knowledge of topics in engineering mathematics – Fourier series, differential calculus, complex numbers, probability and statistics etc. – as typically found in the first year curriculum, and will be continuing to develop his or her understanding of these topics by concurrent second-year studies. It is further supposed that students will have some familiarity with continuous-time signals and systems, including the Laplace transform and s-plane techniques, as a result of prior or concurrent study.

The book has grown out of lectures delivered at the University of Southampton over the last 12 or so years. At first, the material formed part of a third-year course in digital techniques in what was then the Department of Electronics. Over time, much of it has migrated down to second-year level – reflecting the increased importance of the subject which, in turn, reflects the vastly increased availability of low-cost hardware on which to perform digital signal processing. Another manifestation of the technology trend making computing power ever more accessible has been the metamorphosis of my workplace into the Department of Electronics and Computer Science. At present, electronic engineering students in the Department study the majority of the material in this book in a second-semester course, where it forms a half-unit (together with a half-unit of communications theory). This follows a first-semester course in which they are acquainted with continuous-time signals and systems – hence, my assumptions above. There is also some introductory material on signals and systems taught in first year. I have tried, however, to make the book reasonably self-contained, both to suit it to students from other institutions and also in the belief (borne out of painful experience) that a useful understanding of this fascinating subject comes only from repeated exposure to its key principles.

The topic of signals and systems is a notoriously difficult one for beginning students, as a consequence of its mathematical foundations and concepts. It is my belief, and one which has materially shaped this book, that only by grappling with the underlying mathematics can a real understanding be gained. Accordingly, the style of writing attempts to acknowledge students' difficulties and every effort is made to explain the necessary mathematics as it is encountered, but to do so from the point of view of a practical signal-processing engineer. Better mathematicians than I am could doubtless give more rigorous explanations of difficult points, but I am not sure that such explanations would always be useful for engineering students at this level. So I have tried to make accessibility my watch-word and, consequently, rigour may sometimes suffer – although not too badly, I trust.

There is a wealth of excellent reference books on discrete-time signals and systems, but they tend to be voluminous and to make heavy use of mathematical insights and facilities that the student at this level will still be in the process of acquiring. Not infrequently, they claim to be based on a one-semester course of lectures at the author's or authors' institution, but this does leave one wondering just how it is possible to cover such a mass of difficult material in such a short time. In writing this book, my objectives have been more modest: I have tried to cover only essentials. Throughout, I have had a clear goal in mind: namely, that the reader who has mastered the material here will have the knowledge and confidence to approach the standard reference books, understand them, and be able to use them as the basis of further, more advanced study and specialisation.

It is becoming increasing common for introductory books on this subject to include demonstration software. Early on in the (very long) gestation period of the present book, I thought hard about whether to do so here. While, undoubtedly, much can be learned about discrete-time signals and systems by experimentation with carefully designed software, in the end I decided against for two reasons. The first of these was essentially practical. When I started writing, my choice of programming language for demonstration software would have been Pascal. Later it would have been C. Then it might have been C++. This lead me to the belief that the inclusion of such software was a prime way to build obsolescence into the book. Also, we now have easy availability of mathematical software packages at reasonable cost, many of which serve very well to demonstrate the material in this book. Accordingly, the student is encouraged to try using such packages when working through the examples included at the end of each chapter. The second reason was that, as stated above, I am aiming to impart depth of understanding of the key concepts. With this understanding, it then becomes relatively straightforward for the student to write his or her own software. I believe, therefore, that it is well worth encouraging students to test their understanding of the topic of discrete time signals and systems by writing and using their own code. Later on, when mastery is assured, even greater advantage can be taken of software packages – including those specifically tailored to signal processing applications – in confidence that they will be used properly and appropriately.

The book is structured as follows. The first chapter both introduces key concepts of discrete-time signals and systems, and reviews important topics from continuous-time signals and systems theory – in the spirit of keeping the material largely self-contained. Some attention is paid to delta functions and their use in modelling the process of sampling, which links the discrete-time and continuous-time domains. While conceptually simple, the mathematical difficulties in modelling the sampling of a continuous-time signal to yield a discrete-time sequence are revealed – in distinction to most texts which gloss over this rather badly. The sampling theorem is introduced, together with the related issues of aliasing, anti-aliasing and signal reconstruction by interpolation. Treatment of systems is limited throughout the book to the class of linear time-invariant systems. These are introduced, and further important properties defined and developed, in continuous-time terms before making the straightforward generalisation to discrete-time systems. The 'impulse' (or unit-sample) response is then presented as a descriptor of the time-domain properties of a discrete-time system, and the convolution-sum developed as a way of finding the response to arbitrary inputs. Finally in the introductory chapter, Fourier and Laplace descriptions of continuous-time signals are reviewed in preparation for later study.

The second chapter deals with the description of linear time-invariant discrete-time systems in the time domain using difference equations. These are presented as both the discrete-time counterpart to linear differential equations, and as a prescription for implementation. Discrete differentiation and integration (which are, of course, of considerable interest in their own right) are used to exemplify the importance of difference equations as system descriptors. We then explore how the discrete-time counterparts of the methods for solving linear differential equations can be used to find the output of a system for arbitrary inputs, but with some attendant difficulties. The z-transform is then introduced as a means of easing these difficulties.

Treatment of the z-transform commences with a review of the necessary complex variable mathematics – specifically Taylor and Laurent series – to understand the topic properly. The z-transform is seen to have the same sort of relation to a Fourier description of a sampled-data sequence as the Laplace transform has to the Fourier transform in the continuous-time case. After studying the properties of the transform, the system function $H(z)$ is introduced as the z-transform of the impulse response. The geometrical depiction of the system function in terms of a pole-zero diagram is detailed and the specification of the frequency response as the magnitude and phase of the system function is developed. Methods for evaluating the inverse s-transform are then outlined before dealing with the system time response to various inputs – most importantly delta and unit-step functions.

Chapter 4 commences the treatment of discrete-time ('digital') filters by considering the design of recursive filters – i.e. those having an impulse response which is theoretically infinite in time. Unlike the situation when designing analogue filters, there are no synthesis methods which can realise digital filters with arbitrary properties. Hence, design is most usually based on an analogue 'prototype'. Two methods of converting from an analogue prototype to a digital realisation are considered: the bilinear transform and the method of impulse invariance. The first of these attempts to preserve the magnitude response of the prototype while the second attempts to preserve the time response.

Next, the very important topic of the discrete Fourier transform (DFT) is dealt with. This is a Fourier (spectral) descriptor of a sequence (i.e. a discrete-time signal) which is limited in duration. Further, the signal representation is discrete in frequency, which is a pre-requisite for storage and processing in digital hardware. In practice, of course, all signal representations are time-limited; hence, the DFT is of primary importance in the study of discrete-time signals and systems. The DFT and its inverse are defined, and the properties of the transform detailed. The effects of truncation to a finite duration are considered via the topic of windowing. Practical uses of the DFT for interpolation and linear filtering are then outlined.

Based on understanding of the DFT, the design of finite impulse response digital filters is considered. Such filters have important stability properties and can be designed to have linear phase (so that signals are transmitted without phase distortion). The two methods of Fourier series truncation and frequency sampling are detailed.

The practical importance of the DFT owes much to the existence of efficient algorithms for its computation. The class of fast Fourier transform (FFT) algorithms is described in the penultimate chapter, with both decimation-in-time and decimation-in-frequency versions covered. Radix-2 algorithms figure most prominently, but some attention is also paid to composite-N FFTs. This chapter is primarily intended to deepen the reader's appreciation of the DFT. I hope, however, that it can also serve more generally to introduce engineering students to the increasingly important topic (as software continues to gain on hardware as the first-choice implementation technology) of algorithmic efficiency.

In the final chapter, the foundations are laid for further, more advanced study by introducing the topic of random signals. In practice, virtually all signals of interest are in this class. Because random signals are non-deterministic (i.e. they cannot be fully described by a simple mathematical formula), they must be specified by their gross (average) statistical properties. The probability density function, autocorrelation function and power spectrum of a random signal are all dealt with, before turning attention to characterising the relation between two such signals using the joint probability density function, cross-correlation function and cross power spectrum. These cross measures assume particular importance when the two signals of interest form the input to and output from a discrete-time system. The book ends with an introduction to the problem of estimating these signal measures from finite-length sequences.

The mechanics of book production have altered markedly in recent years. I am always surprised these days to see new books come out with acknowledgments paid to whoever typed the manuscript. Widespread availability of cheap (or even free!) desktop publishing or typesetting software makes it very attractive – if not actually mandatory – for authors to do their own text entry. The high degree of interaction and involvement which this promotes between the writer and his or her material means that, personally, I could countenance no other way of working. Thus, it was natural for me to produce this book in camera-ready form. In this, I was encouraged by my commissioning editor at Chapman and Hall, Dave Hatter. (At this point, I must record my debt to Dave for his extreme patience throughout the inordinately long period it took me to finish the book. His belief never wavered that one day I really would present him with final camera-ready copy, and for that I thank him.) So, for those of a technical bent, I give some de tails of the software and hardware used. The text is set using Leslie Lamport's LATEX – a collection of macros for Donald Knuth's TEX typesetting software. A multitude of versions of the software and computational platforms have been used during writing, but final text preparation was done using LATEX Version 2.09 on a Hewlett-Packard 9000/710. I chose Times Roman for the text typeface both for its unrivalled availability but also in view of Mark Hengesbaugh's assertion in *Typography for Desktop Publishing* that "Roman typefaces are, in fact, the best text typefaces". Avalon (Avante Garde) is especially good for for presentational material, and so was initially selected as the typeface for the illustrations. At first, I used two packages to prepare these: Lotus 1-2-3 Release 4.1 for graphing and CorelDraw Version 3.0 for other diagrams. These were chosen because of their advertised ability to work together via Window's object linking and embedding (OLE) facility. The intention was to have a consistent presentation standard by using only CorelDraw for final output. I must state, however, that (for whatever reason) I never managed to have them working in concert and my hopes for consistency went sadly unrealised. Worse still, Lotus 1-2-3 did not produce histograms of appropriate quality so that yet another package (Harvard Graphics Version 2.10) was used for these. Avalon was not available in this package, and Executive typeface was selected. The camera-ready copy has been produced on a Hewlett Packard 4M Plus LaserJet with Resolution Enhancement PCL5 at 600 dots per inch.

Finally, turning to less mundane matters, I wish to acknowledge the debt I owe to the small band of gifted teachers who first introduced me to the topic of discrete-time signals and systems when I was a student in the Department of Electrical Engineering at Imperial College, and instilled in me some of their enthusiasm for scholarship. Foremost among these is Bruce Sayers, but additional thanks must go to Dick Kitney, Don Monro and Paul Lynn. Fellow students Paul Cheung and John Branch also taught me a great deal; they were always ready to discuss my problems of understanding and were a real source of support. John Branch's tragically early death at the age of 27 was an ob ject lesson for me: time is short for us to achieve all that we may want to. Since I have been at Southampton, I have grown to admire very much Joe Hammond's approach to the teaching of digital signal processing – based on insightful analogies and clear explanations, backed up by boundless enthusiasm. My appreciation of the subject has been heightened by many informal discussions with colleagues Mark Nixon, Christine Shadle, Sean Smith, John Carter and David Nunn. Mark, in particular, read earlier versions of the manuscript and commented helpfully. It is my hope that this book will, in some small way, take what I learned from all of them and use it to inspire a new generation of students to master this fascinating and rewarding topic.

R.I. Damper
Chandler's Ford
April 1995

1

Introduction

This book is concerned with the basic theory that underlies the processing of signals by digital hardware. So, what precisely is a signal? One dictionary defines it as "an intelligible sign conveying information ... especially ... at a distance". This is clearly a very wide definition indeed, and its very broadness is a pointer to the tremendous practical importance of the subject we will be studying. Some representative examples of signals might be:

- the infra-red signal used to switch or adjust controls remotely on a domestic television and/or video cassette recorder;
- the sound pressure wave of human speech, and the electrical representation of that speech transmitted along a telephone cable;
- a sequence of electrical pulses in a telegraphic system, where the information conveyed is some textual message represented in, say, Morse code;
- the sequence of midday dollar exchange rates on the London stock exchange over some period of days.

It should be clear from these few simple examples that the notion of a signal is pervasive, occurring in areas as diverse as radio communications, biomedical engineering, telematics and control, financial forecasting, remote sensing, sonar, seismology and many others beside.

What then does it mean to process a signal? Here, again, the possibilities are myriad. If a speech signal, for instance, is to be transmitted over a distance, then we will be concerned with representing the information in the signal compactly as this affects the economics of the communication system. Thus, we may process the signal to remove redundancy or to allow efficient coding. During transmission, the speech signal may become contaminated by spurious noise so that, at the receiving end, we might process it to improve its quality or intelligibility. In many signal-processing applications, we will be trying to extract automatically from the signal important information and make it explicit. For instance, can we predict likely catastrophic failure by monitoring sound emission from an engine, or the electrical activity of a patient's heart? What does a pattern of reflected seismic waves tell us about the possible presence of pockets of oil or other valuable minerals in the sub-strata? Do observed water-borne acoustic waves originate from a submarine or a car ferry? In other words, how do we *analyse* a signal in order to learn something about the source that produced it, or the transmission channel though which it has passed? In addition to analysis, we may too wish to design systems to *synthesise* signals from some high-level description. For instance, the technology of synthesising speech from textual input is starting to assume commercial importance. Overall, the sheer potential of engineering systems for processing signals digitally is simply enormous. Additionally, many interesting and important problems can be framed as *system identification*. That is, given an input signal to a system and an output signal from it, what can be inferred about the system itself?

The reader just beginning his or her study of signals and systems may wonder why these two concepts seem invariably to be dealt with together, almost as an indivisible whole. There are at least two ways of answering this question. First, a system in our terms is only of interest in as much as it processes signals, and signals are only meaningful to the extent that we can process them by some system to enhance, reveal or transform the information they contain. A second reason will, however, emerge early in our study if indeed you are not already familiar with it. An important class of systems (indeed the only ones we will be concerned with in this introductory text) is the so-called linear time-invariant (LTI) class – see section 1.8 below. Such systems can be uniquely described by a particular signal called the *impulse response* which, as the name suggests, is the system's output when excited by a defined impulsive input. The fact that LTI systems can be identified and described by a rather special signal is an additional and powerful reason for treating the two together.

Most of the signals we have so far mentioned are essentially one-dimensional (1D), being functions of amplitude versus the single variable time. This is far from a necessary restriction; the independent variable need not be time. Also, signals of two or more dimensions are of great practical interest and importance. An obvious example is a two-dimensional (2D) image in which intensity varies as a function of, say, horizontal and vertical spatial coordinates. A time sequence of such 2D images would constitute a three-dimensional signal. Generally speaking, the processing of 2D or higher multidimensional signals uses techniques that are extensions of 1D signal processing. Hence, for the purposes of this book, we simplify the treatment by restricting ourselves to 1D signals and signal-processing systems exclusively.

This introductory chapter proceeds as follows. First, we contrast discrete-time (DT) and continuous-time (CT) signals before looking in some depth at delta functions which are important to the study and signals and systems. Both CT and DT versions of delta function exist and are compared. We then consider the conversion of CT signals to DT form by sampling, as well as the closely-related process of analogue-to-digital conversion, and the problems which can arise in this. These problems can be minimised by filtering the CT (analogue) signal before sampling. Subsequently, we look at the process of reconstructing, or interpolating, the original CT signal from its samples. The concepts of power and energy are then extended to apply to DT signals and a classification of DT signals made according to whether average power and/or total energy are finite or infinite. We then review the properties of a particular class of signal-processing systems – the linear, time-invariant (LTI) systems. Since a considerable proportion of this book is devoted to describing DT signals in terms of a (complex) frequency domain, we next review Fourier descriptions of continuous-time signals with which the reader is assumed to have at least some prior acquaintance. We then briefly deal with the Laplace transform as a generalisation of the Fourier transform. This is intended to lead in to later study of the z-transform which is of major importance in describing discrete-time signals and systems.

1.1 Continuous- and discrete-time signals

Many of the example signals discussed above are electrical in nature (as well as usually being one-dimensional functions of time) but, equally, many are not. Rapid advances in the art and technology of analogue electronics in the mid-years of this century made it increasingly attractive to process information electrically. Thus, an important part of electronic engineering has been the development of sensors and transducers to convert non-electrical signals, such as pressure or temperature variations, into electrical form. The latter half of the century has, however, seen phenomenal growth in the capabilities of digital electronic technology which have relegated analogue techniques to a position of secondary importance. Digital systems have vastly better noise immunity and tolerance to faults and drift in component values, as well as being far easier to design, manufacture, test and maintain. Even more importantly, however, modern digital hardware is now characterised by software programmability which offers the system designer unrivaled flexibility in developing solutions to a variety of signal-processing problems.

Unlike analogue systems, however, digital hardware can only process data which are discrete. As very many real-world signals of interest are essentially continuous in time, some means is necessary for converting continuous-time (CT) signals to discrete-time (DT) form. This is done by *sampling* – the topic of section 1.3 below. Sampling restricts the signal to pre-specified instants of time: in this book, these instants will always be regularly spaced by the (constant) sampling interval T. The reciprocal of T is the *sampling frequency* or *sample rate*, f_s. Thus, given the CT signal $x(t)$, sampling yields a sequence of values $x(nT)$ where n is an integer. This sequence is the DT signal: hence, we will use the terms *sequence* and (DT) *signal* interchangeably in this book. For practical processing, $x(nT)$ will obviously need to be finite in length. This is achieved by limiting – or *windowing* – the CT signal in time. Windowing is an important topic which we will deal with in some detail later.

Alternative and entirely equivalent notations to $x(nT)$ are $x(n)$ and x_n with the sampling period parameter T simply implied. We will most often use $x(n)$, but will also use the other notations liberally, and as appropriate to reduce confusion about our meaning. Strictly, $x(n)$ refers to a particular value of the sequence – that at time index n – rather than the entire sequence. The latter is correctly denoted by the set $\{x(n)\}$. However, the tradition in the subject is to ignore this distinction and to use $x(n)$ to denote both a particular sample value and the entire sequence. It is usually clear from context if the n index is specific or general. We will, however, need to make this distinction implicit at one later point (chapter 8).

Strictly, the term "discrete-time" implies that only the time variable has been quantised, but for digital processing the amplitude dimension must also be discretised. The process of obtaining such a representation from a CT signal is called *analogue-to-digital conversion*. The implications of amplitude quantisation – representing an effectively continuous variable by a finite-length digital code-word – are important but essentially practical in nature. In order to keep basic conceptual issues to the fore, we will ignore the topic of finite word-length representations in this book, leaving it for later study in a more advanced course. This justifies the wording *discrete-time signals and systems* in our title rather than, for example, *digital signal processing*.

1.2 DELTA AND STEP FUNCTIONS

The notion of a sampled-data sequence which exists only at specific instants of time ($t = nT$) is an intuitively attractive one, but is not without its conceptual and mathematical difficulties. The basis of the difficulty is that, although well-defined, an 'instant' has infinitesimally small extent in time. The concept has similarities with other abstractions in physics, such as a point charge, a point load on a beam, the voltage induced when an inductive circuit is broken in 'zero' time or position/momentum states of a wave-function in quantum mechanics. Situations such as these can be usefully modelled using the so-called *delta function*, $\delta(t)$, introduced by the brilliant British physicist Paul Dirac in his pioneering work on quantum mechanics. (Actually, it is not a 'function' in the strict sense but we will ignore this here.)

The Dirac delta function has some rather strange properties. It is a function of continuous time, t, but only exists at time t_0:

$$\delta(t - t_0) = 0 \qquad t \neq t_0$$

At this instant, it is infinitely narrow and infinite in height but has unit area. Hence, its definition is completed by specifying its behaviour under the integral sign, when multiplied by some general function $f(t)$:

$$\int_{t_1}^{t_2} f(t)\delta(t - t_0)dt = f(t_0) \qquad t_1 < t_0 < t_2 \tag{1.1}$$

so that, if $f(t) = 1$ over all time:

$$\int_{-\infty}^{\infty} \delta(t - t_0)dt = 1 \tag{1.2}$$

thus satisfying the unit-area property.

An alternative definition of the Dirac delta takes it to be the limiting case of a positive rectangular pulse centred at t_0, having width ϵ and height $1/\epsilon$, as ϵ tends to zero:

$$\delta(t - t_0) = \lim_{\epsilon \to 0} p(t, \epsilon)$$

$$\text{where} \quad p(t, \epsilon) = \begin{cases} 0 & t \leq t_0 - \epsilon/2 \\ 1/\epsilon & t_0 - \epsilon/2 < t < t_0 + \epsilon/2 \\ 0 & t \geq t_0 + \epsilon/2 \end{cases}$$

This is depicted in figure 1.1 for the case of $t_0 = 0$.

Although we have used a rectangular pulse here, it is in fact possible to use a variety of functions to define the Dirac delta in similar fashion.

The importance of equation (1.1) is that it can be used to extract from a CT function, $f(t)$, the specific value that exists at time t_0 (or nT) (figure 1.2). This equation defines the so-called *sifting property* of the Dirac delta function and suggests potential applicability to the process of sampling a CT signal to produce a DT version.

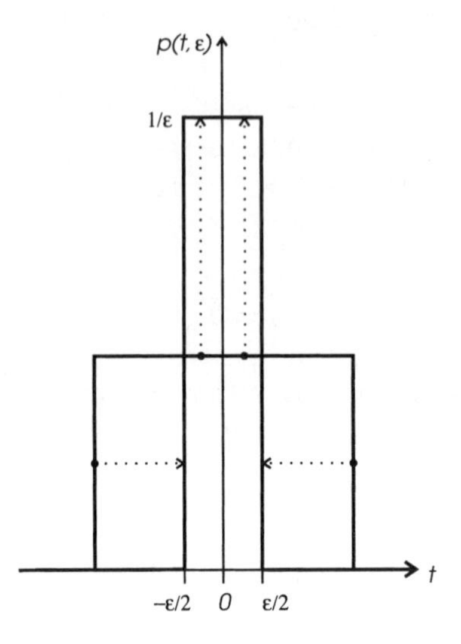

Fig. 1.1 The Dirac delta function can be defined in terms of the limit of a rectangular pulse of width ϵ and height $1/\epsilon$ as ϵ tends to zero.

However, the Dirac delta, $\delta(t)$, is a function of *continuous* time in that it is, in principle, defined over all time. When working in the domain of discrete time, it is helpful to introduce a similar function which is only defined at discrete instants, nT say. This new function has unit amplitude (rather than unit area) only when the time index n takes some particular value k, and is called the unit-sample sequence or the Kronecker delta function :

$$\delta(n-k) = \begin{cases} 1 & n = k \\ 0 & \text{otherwise} \end{cases}$$

Here, of course, the variables n and k are discrete. The Kronecker delta is sometimes written equivalently as δ_{nk}. When the delta function is positioned at the origin ($k = 0$), we will simply write δ_n.

Considering equations (1.1) and (1.2) and recognising that t in the integrations is a dummy variable (which can therefore be replaced by x, say), let us define a new CT function:

$$u(t-t_0) = \begin{cases} \int_{-\infty}^{t} \delta(x-t_0)dx = 0 & t < t_0 \\ \int_{-\infty}^{t} \delta(x-t_0)dx = 1 & t \geq t_0 \end{cases} \tag{1.3}$$

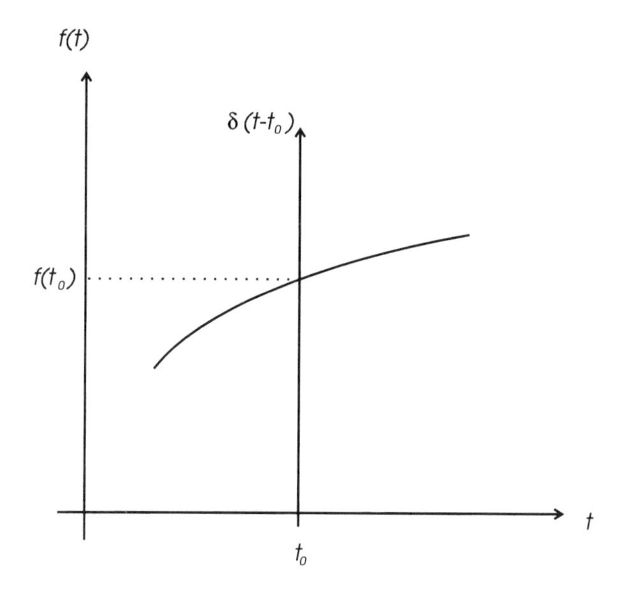

Fig. 1.2 The sifting property of the Dirac delta function $\delta(t - t_0)$ can be used to extract from a CT function the specific value of the function at time $t = t_0$ by integrating over time. The actual height of the delta function at $t = t_0$ is undefined.

This is the unit-step function which, by this definition, is the derivative with respect to time of the Dirac delta function. It is a simple matter to define a discrete-time version, the unit-step sequence, as:

$$u(n-k) = \begin{cases} 0 & n < k \\ 1 & n \geq k \end{cases}$$

Again, as we did for the unit-sample (Kronecker delta), we will write u_n for the unit-step sequence positioned at the origin ($k = 0$).

Clearly, because the unit-step sequence exists only at discrete times, it can be expressed just as well in terms of unit-sample sequences as:

$$\begin{aligned} u(n-k) &= \delta(n-k) + \delta(n - \{k+1\}) + \delta(n - \{k+2\}) + \cdots \\ &= \sum_{m=0}^{\infty} \delta(n - \{k+m\}) \end{aligned}$$

$$= \sum_{r=k}^{\infty} \delta_{nr} \qquad \text{with } r = k + m \tag{1.4}$$

It is sometimes useful to think of the Kronecker delta as the difference between two unit-step sequences offset by a single sample period:

$$\delta_{nk} = u(n - k) - u(n - \{k + 1\}) \tag{1.5}$$

which follows directly from equation (1.4). Interestingly, given that the Dirac delta is the derivative of the (CT) unit-step function, this differencing operation can be thought of as the DT counterpart of differentiation – a point which we take up in the next chapter.

1.3 SAMPLING

The situation which concerns here is as follows. We have a signal $x(t)$ which is a continuous function of time, and we wish to process it by digital computer (or perhaps more specialised digital hardware). This requires us to find a set of sample values $x(nT)$ (or, more simply, $x(n)$ with the sampling period implied) such that:

$$x(n) = x(t)|_{t=nT}$$

This is depicted in figure 1.3, where the sampler can be visualised as an on/off switch which is ideal in the sense that it makes and breaks instantaneously at times $t = nT$.

Now, an arbitrary sequence $x(n)$ of regularly-spaced values is well described as a sum of scaled and delayed unit-samples:

$$x(n) = \sum_{k=-\infty}^{\infty} x(k)\delta(n - k) \tag{1.6}$$

This, of course, is a more general statement of the very formulation employed in writing equation (1.4) above.

In practice, signals are not of infinite extent in time and there is some value n_0 of n such that:

$$x(n) = 0 \text{ for all } n < n_0$$

and $x(n)$ is then called a *right-sided* sequence. Thus, for a right-sided sequence:

$$x(n) = \sum_{k=n_0}^{\infty} x(k)\delta(n - k)$$

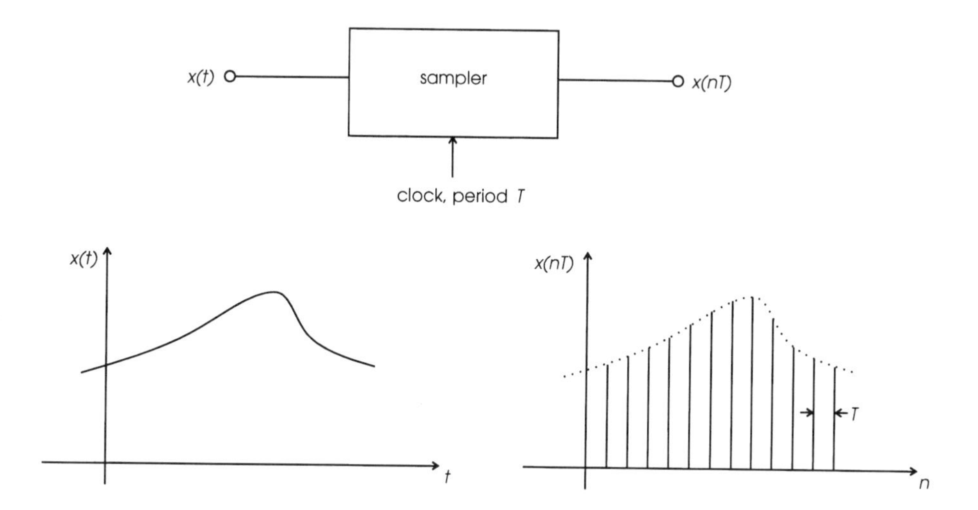

Fig. 1.3 Sampling of a continuous time signal $x(t)$ to yield the sequence $x(nT)$. The sampler is visualised as an ideal on/off switch.

The sampling problem is then to derive a sequence in the form (1.6) from $x(t)$. One idealised way to do this, referred to as *instantaneous sampling*, uses the Dirac delta function as a 'bridge' between the continuous- and discrete-time domains.

Consider the so-called *comb* function (figure 1.4):

$$c_T(t) = \sum_{n=-\infty}^{\infty} \delta(t - nT) \tag{1.7}$$

We define a new function $x^*(t)$ where:

$$x^*(t) = x(t)c_T(t)$$

Let us assume for simplicity that $x(t) = 0$ for $t < 0$ so that the sampled data sequence is right-sided and:

$$x^*(t) = \sum_{n=0}^{\infty} x(t)\delta(t - nT) \tag{1.8}$$

By virtue of the appearance of the Dirac delta function on the right-hand side, with its rather peculiar properties, equation (1.8) is actually something of an abstraction. It has an obvious similarity to equation (1.6) describing a DT sequence. However, $x^*(t)$ must be a CT function since it involves $x(t)$ and $\delta(t - nT)$ only, yet because $\delta(t - nT)$

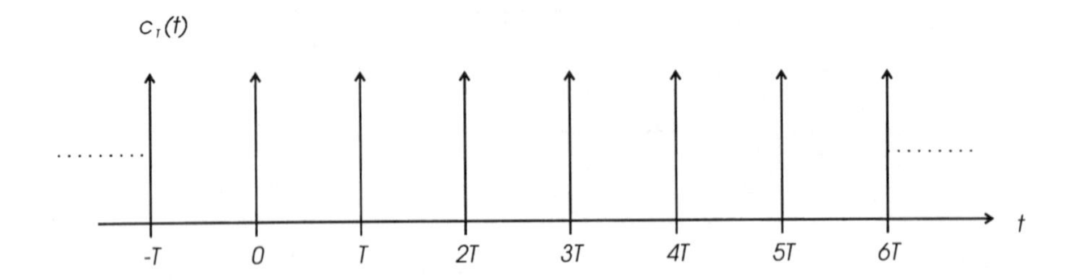

Fig. 1.4 The comb function $c_T(t)$ consists of a sum of Dirac delta functions at integer delays nT. The height of each delta function is indeterminate, being defined by the behaviour under the integral sign.

only exists when $t = nT$ so does $x^*(t)$. Thus, there are some difficulties of interpretation in the use of equation (1.8). These are very much eased if we are in a position to integrate $x^*(t)$, since we can then invoke the sifting property. This allows us to think of the sampled signal as a sequence of impulses whose *areas* are equal to the values of $x(t)$ at times $t = nT$. We will see an example of this later in chapter 3.

Another idealised form of sampling is called *natural* sampling. In this case, we multiply $x(t)$ by a train of rectangular pulses of unit height and width τ, where τ is thought of as small relative to the rate of change of $x(t)$. In the limit $\tau \to 0$ therefore, the individual pulses become (unit-sample) Kronecker deltas and $x(n)$ is well-described by equation (1.6).

In this book, we will not consider hardware for analogue-to-digital conversion in any detail. As well as a *practical* (rather than idealised) sampler as in figure 1.3, an analogue-to-digital converter also includes some form of amplitude quantiser as well as a coder to convert the quantised amplitude into binary form. Unlike the sampling process itself, which is theoretically reversible (see below), quantisation is irreversible.

1.4 ALIASING AND THE SAMPLING THEOREM

Consider a CT cosine wave of amplitude A, frequency f and phase θ:

$$x(t) = A\cos(2\pi ft + \theta)$$

Suppose we now sample this with $f_s = Nf$ or $T = 1/Nf$, i.e. so that there are N samples per period. The corresponding discrete-time sequence is:

$$x(n) = A\cos(2\pi ft + \theta)|_{t=nT}$$

$$= A\cos\left(\frac{2\pi fn}{Nf}+\theta\right)$$

$$= A\cos\left(\frac{2\pi n}{N}+\theta\right)$$

This, however, indicates a problem since the cosine function takes the same value for many different arguments. In particular, for any ϕ:

$$\begin{aligned}\cos(\phi) &= \cos(-\phi)\\ \cos(\phi) &= \cos(\phi+2\pi kn) \quad \text{for integer } k \text{ and } n\\ \text{and so } \cos(\phi) &= \cos(2\pi kn \pm \phi)\end{aligned}$$

Setting $A = 1$ and $\theta = 0$ for clarity:

$$x(n) = \cos\left(\frac{2\pi n}{N}\right) = \cos\left(2\pi kn \pm \frac{2\pi n}{N}\right) \tag{1.9}$$

Equation (1.9) indicates that, given a sequence $x(n)$ produced by sampling from a CT cosine (or sine) wave, there is ambiguity about the frequency f of the underlying CT signal. Since, by Fourier's theorem, any signal can be represented by a sum of sine and cosine waves of appropriate phase, this problem of ambiguity is an important and general one.

As an illustration of the problem, suppose the CT cosinusoid $x_0(t)$ has frequency $f_0 = 100$ Hz and is sampled with $f_s = 400$ Hz ($N = 4$) so that:

$$\begin{aligned}x_0(n) &= \cos\left(\frac{2\pi 100n}{400}\right)\\ &= \cos\left(\frac{2\pi n}{4}\right)\end{aligned}$$

Now consider a second cosine wave $x_1(t)$ of frequency $f_1 = 300$ Hz sampled at the same rate of $f_s = 400$ Hz:

$$\begin{aligned}x_1(n) &= \cos\left(\frac{2\pi 300n}{400}\right)\\ &= \cos\left(\frac{6\pi n}{4}\right)\end{aligned}$$

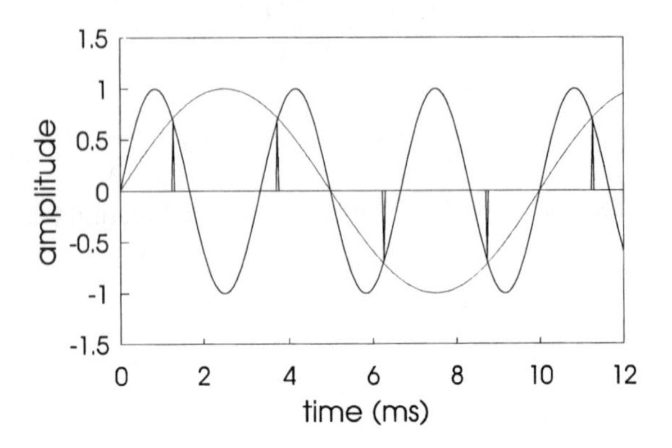

Fig. 1.5 A 300 Hz sinusoid sampled at 400 Hz (i.e. below the Nyquist rate) aliases a lower-frequency (100 Hz) wave.

$$= \cos\left(2\pi n - \frac{2\pi n}{4}\right)$$

By equation (1.9), taking the negative sign and with $k = 1$, $x_1(n)$ and $x_0(n)$ are identical. So, given this sequence, it is not possible to tell whether it arose by sampling a 100 Hz or a 300 Hz wave. Thus, we say that $x_1(t)$ when sampled at 400 Hz *aliases* a lower-frequency (co)sinusoid (of frequency 100 Hz) as illustrated in figure 1.5. Indeed, the same sequence would arise by sampling waves of higher frequency, with $k > 1$ and/or taking the positive sign in equation (1.9). The ambiguity can be resolved if somehow we know always to take the lowest frequency, i.e. by interpreting the sequence considered here as a sampled-data representation of $x_0(t)$. This leads us to the very important *Nyquist sampling theorem* which at this stage we will state without formal proof:

> If the highest frequency contained in a CT signal $x(t)$ is $f_{max} = B$, then $x(t)$ can be recovered exactly from its sample values $x(nT)$ provided the sampling frequency $f_s = 1/T$ is such that $f_s > 2f_{max}$.

Throughout the remainder of the book, we will have frequent recourse to this theorem and ample opportunity to verify it.

You may very well have seen aliasing before on film or television, when the spoked wheels of a stage-coach appear to rotate backwards. This is because the video frame rate *under-samples* the rotation of the spokes.

Applied to a simple sine or cosine wave, the Nyquist theorem states that we need at least two samples per period in order to sample the wave *properly*, i.e. so that it can be recovered exactly from the sample values. As figure 1.5 shows, the 300 Hz sinusoid does not satisfy this *Nyquist criterion* while the 100 Hz wave has $N = 4$ samples per period.

Consider equation (1.9) with $k = 1$ and taking the negative sign:

$$\cos\left(\frac{2\pi n}{N}\right) = \cos\left(2\pi n - \frac{2\pi n}{N}\right)$$

where the sequence on the left-hand side has been obtained by properly sampling a wave of frequency f_0.

But $f_s = N f_0$, so that:

$$\begin{aligned}\cos\left(\frac{2\pi f_0 n}{f_s}\right) &= \cos\left(\frac{2\pi f_s n}{f_s} - \frac{2\pi f_0 n}{f_s}\right) \\ &= \cos\left(\frac{2\pi (f_s - f_0) n}{f_s}\right)\end{aligned}$$

where the right-hand side corresponds to a wave of frequency f_1 which aliases a wave of frequency f_0 such that:

$$f_1 = f_s - f_0 \tag{1.10}$$

It follows immediately that the arithmetic mean of f_0 and f_1 is:

$$\frac{f_0 + f_1}{2} = \frac{f_s}{2}$$

This is called the *Nyquist* or *folding* frequency.

From equation (1.10), it is obvious that $f_1 < f_s$. However, because there is aliasing, it must also be the case that $f_s < 2f_1$. Hence:

$$\frac{f_s}{2} < f_1 < f_s$$

and f_1 lies between the folding and the sampling frequencies.

Returning to equation (1.9) with $k = 1$ but taking the positive sign:

$$\cos\left(\frac{2\pi n}{N}\right) = \cos\left(2\pi n + \frac{2\pi n}{N}\right)$$

$$\text{and} \quad \cos\left(\frac{2\pi f_0 n}{f_s}\right) = \cos\left(\frac{2\pi f_s n}{f_s} + \frac{2\pi f_0 n}{f_s}\right)$$

$$= \cos\left(\frac{2\pi(f_s + f_0)n}{f_s}\right)$$

Here, the right-hand side corresponds to a wave of frequency $f_2 = f_s + f_0$ which aliases f_0.

Continuing in this way for increasing values of k and considering both positive and negative signs in equation (1.9) we see that, in general, the aliasing frequencies are $kf_s \pm f_0$.

1.5 ANTI-ALIASING

We stated above that the inherent ambiguity in interpreting a sampled-data sequence is resolved if we know to take the lowest possible (positive) frequency from the set $kf_s \pm f_0$, i.e. f_0. Given the Nyquist theorem, therefore, we must ensure when sampling a CT signal $x(t)$ that there is no frequency component in $x(t)$ greater than $f_s/2$. In other words, we want the CT signal to be *band-limited* to frequency $B = f_s/2$.

One obvious way to band-limit a signal is to low-pass filter it. Prior to sampling, therefore, it is standard practice to filter the CT signal so as to reduce problems of aliasing. Hence, the process is called *anti-aliasing*. A number of important points should be made about the process.

First, it is clear that anti-alias filtering produces a different (analogue) signal. Accordingly, one must set the *cut-off* of the filter and the sample rate so as to avoid losing too much important information in the high frequencies. In general, there is no simple way to decide what these should be for any particular signal. Rather, one relies on knowledge of the signal and/or intuition. Suppose, for instance, we are to process a speech signal. Since the upper frequency limit of the human hearing system is approximately 20 kHz in young people falling to 15 kHz or less with increasing age, there would probably be little point in retaining frequency information much above about 12 kHz when processing speech.

Second, anti-aliasing is done in the continuous-time (analogue) domain and analogue filters do not have infinite (or even usually very sharp) fall-off. Thus, filtering does not remove high-frequency components entirely, but only attenuates them. Of course, our signal conditioning and sampling hardware does not have infinite sensitivity and resolution, so that it suffices to reduce the high-frequency components to something approximating the noise levels of the hardware.

Analogue low-pass filters are conveniently described by their -3 dB-down or cut-off frequency, f_c. It seems to be very tempting to students to interpret the sampling theorem as saying that f_s should be set at twice f_c on the (mistaken) assumption that the filter will band-limit $x(t)$ to $B = f_c$. This is absolutely wrong. Not only is -3 dB-down nowhere near the noise level of any reasonable, practical signal conditioning hardware (-50 to -60 dB is more realistic), but the theorem sets $2B$ as the theoretical *absolute minimum* for the sample rate. When using an anti-alias filter with only modest fall-off (say 12 or 18 dB/octave), something like 4 or 5 times f_c is a practical choice.

1.6 INTERPOLATION

According to the sampling theorem, $x(t)$ can – if properly sampled – be recovered exactly from the samples $x(nT)$ (provided they have not been quantised in amplitude for digital representation and/or storage). The reconstruction of $x(t)$ from its samples is called *interpolation* since the process involves obtaining values of $x(t)$ at times intermediate between the sample points $t = nT$. We will leave for later study (section 5.9) the question of *ideal interpolation* which yields $x(t)$ exactly for any value of t.

In a practical situation, the sample values will be stored in digital memory, and so will be quantised in amplitude. The reconstruction process, which can never be theoretically exact, is called *digital-to-analogue* conversion. Generally, any attempt at anything like (unachievable) ideal interpolation is too complex for practical conversion and considerably simpler methods are therefore used.

One such method is called *zero-order hold*, illustrated in figure 1.6. This simply takes the value of the previous sample and maintains it over the sampling interval to produce a 'staircase' approximation to $x(t)$. Another fairly obvious approach – although trickier to implement – is linear interpolation, where a straight line is drawn between the sample values.

The process of reconstructing an analogue signal from its DT samples can be expressed mathematically using an interpolation function, $\text{int}(n, t)$. This is a CT function usually, but not always, centred on the nth sample. Assuming a right-sided sequence with $n_0 = 0$, the reconstruction $x_r(t)$ of $x(t)$ is given as:

$$x_r(t) = \sum_{n=0}^{\infty} x(nT)\text{int}(n, t) \tag{1.11}$$

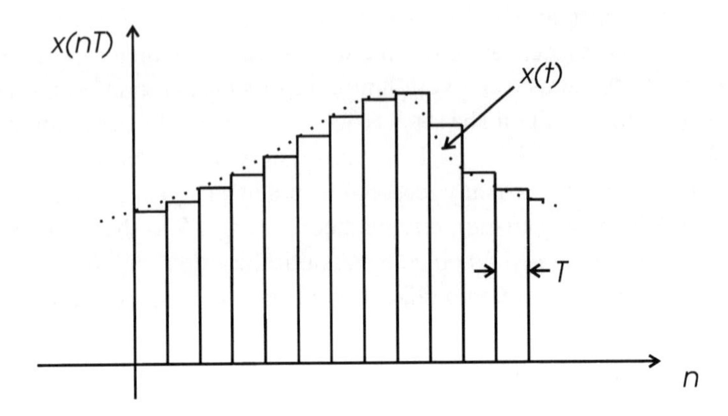

Fig. 1.6 Zero-order hold digital-to-analogue conversion produces a 'staircase' approximation to $x(t)$.

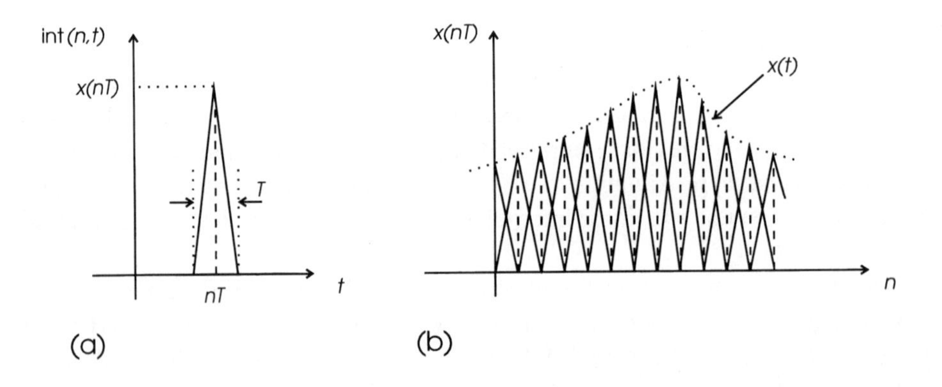

Fig. 1.7 The triangular interpolation function in (a) is used as in (b) to give a linear interpolation of the sequence $x(nT)$.

It can be shown (exercise 1.4) that linear interpolation involves the use of a triangular interpolation function, as depicted in figure 1.7.

1.7 POWER AND ENERGY OF DT SIGNALS

For a (real) discrete-time signal, the concepts of power and energy (which are so useful for describing CT signals) can be made concrete by considering the signal to be a sampled voltage or current wave which is applied to a 1 Ω resistor.

Initially, let us take a continuous-time signal, $x(t)$. If this signal is periodic, then the *average power* can be found from the mean-square value evaluated over a long observation interval, τ, as:

$$P_{av} = \lim_{\tau \to \infty} \frac{1}{\tau} \int_{-\tau/2}^{\tau/2} x^2(t)dt \tag{1.12}$$

For instance, in the case of a sine wave $x(t) = V \sin \omega t$:

$$\begin{aligned}
P_{av} &= \lim_{\tau \to \infty} \frac{1}{\tau} \int_{-\tau/2}^{\tau/2} V^2 \sin^2 \omega t dt \\
&= \lim_{\tau \to \infty} \frac{V^2}{2\tau} \int_{-\tau/2}^{\tau/2} (1 - \cos 2\omega t)dt \\
&= \lim_{\tau \to \infty} \frac{V^2}{2\tau} \left([t]_{-\tau/2}^{\tau/2} - \left[\frac{\sin 2\omega t}{2\omega} \right]_{-\tau/2}^{\tau/2} \right) \\
&= \lim_{\tau \to \infty} \frac{V^2}{2\tau} \left(\tau - \left[\frac{\sin\omega\tau - \sin(-\omega\tau)}{2\omega} \right] \right) \\
&= \frac{V^2}{2} - \lim_{\tau \to \infty} \frac{V^2}{2\tau} \left(\frac{\sin \omega\tau}{\omega} \right) \\
&= \frac{V^2}{2}
\end{aligned}$$

which is the mean-square value.

However, not all signals have a non-zero average power. To see this, consider an aperiodic signal such as a decaying exponential:

$$x(t) = \begin{cases} Ve^{-\alpha t} & t \geq 0 \\ 0 & t < 0 \end{cases}$$

with $\alpha > 0$. For this signal:

$$\begin{aligned} P_{av} &= \lim_{\tau\to\infty} \frac{1}{\tau} \int_0^{\tau/2} V^2 e^{-2\alpha t} dt \\ &= \lim_{\tau\to\infty} \frac{V^2}{\tau} \left[\frac{-e^{-2\alpha\tau}}{2\alpha} \right]_0^{\tau/2} \\ &= \lim_{\tau\to\infty} \frac{V^2}{2\alpha\tau} \left(1 - e^{-\alpha\tau}\right) \\ &= 0 \end{aligned}$$

While this decaying exponential signal has zero average power, its total energy is finite:

$$\begin{aligned} E_{tot} &= \lim_{\tau\to\infty} \int_0^{\tau/2} V^2 e^{-2\alpha t} dt \\ &= \lim_{\tau\to\infty} \frac{V^2}{2\alpha} \left(1 - e^{-\alpha\tau}\right) \\ &= \frac{V^2}{2\alpha} \end{aligned}$$

The important distinction which these two examples illustrate is that between *finite-power* and *finite-energy* signals.

As the name suggests, finite-power signals have non-zero average power but infinite total energy. Periodic signals and some random signals extending over all time are in this category. By contrast, finite-energy signals have zero average power but, of course, finite total energy. This class includes transients and other time-limited signals.

The distinction extends naturally from CT to DT signals. For a finite-power DT signal:

$$P_{av} = \lim_{N\to\infty} \frac{1}{N} \sum_{n=-N/2}^{N/2-1} x^2(n) \neq 0 \tag{1.13}$$

while for a finite-energy DT signal:

$$E_{tot} = \sum_{n=-\infty}^{\infty} x^2(n) < \infty$$

In this case, we say that $x^2(n)$ is *absolutely summable.*

As we shall see, the distinction between finite-power and finite-energy signals is a fundamental one with many important implications.

1.8 LINEAR TIME-INVARIANT SYSTEMS

In this book, we are concerned with DT signals and systems for processing such signals. These systems will generally be intended to enhance, reveal or transform the information contained in the input signal or signals. A large and very important class of discrete-time systems are those known as *linear.* The importance of these systems derives from the fact that powerful (linear) mathematical techniques can be used to describe them. Further, linear CT systems based on these mathematical descriptions make excellent models of many real, physical systems. Linear DT systems can, therefore, often be usefully regarded as discrete-time counterparts of such CT models.

1.8.1 Linearity

As just indicated, the term *linear* derives from the mathematical description. Considering a CT system, this description is in terms of an nth-order linear differential equation with constant coefficients:

$$c_n\frac{d^n y}{dt^n}+c_{n-1}\frac{d^{n-1}y}{dt^{n-1}}+\cdots+c_1\frac{dy}{dt}+c_0 y = R(t)$$

or

$$c_n\frac{d^n y}{dt^n}+c_{n-1}\frac{d^{n-1}y}{dt^{n-1}}+\cdots+c_1\frac{dy}{dt} = -c_0 y+R(t) \qquad (1.14)$$

That is, the expression on the right-hand side is linear in y. In system terms, y can be taken to represent the 'output' with the system input appearing in $R(t)$, which is called the *forcing function.* In the general case, the forcing function can include time derivatives of the input additionally but, for simplicity, we do not show this explicitly here. (We should, however, note that the derivatives of the input can affect the order of the equation. If an mth derivative of the input appears, and $m > n$, then the equation is mth order.)

For instance, the instantaneous voltage v across a capacitor C is related to the applied current i as the first-order equation:

$$i = C\frac{dv}{dt} \qquad \begin{cases} c_n = 0 \quad n \neq 1 \\ c_1 = 1 \\ R(t) = i/C \end{cases}$$

confirming the familiar fact that the (ideal) capacitor is a linear circuit element.

A common pitfall is to assume that 'linear' means there is a straight-line relationship between the system input and output, i and v respectively in this simple example of a capacitor. The reader should note carefully from the above that this is an over-simplification.

If we 'solve' the simple differential equation above, we obtain:

$$v = \frac{1}{C}\int i dt + k$$

where the constant of integration k denotes the possibility that the capacitor may be initially charged at the instant the current is first applied. Leaving aside this possibility for the moment, the right-hand side of the equation can be viewed as representing a *linear operator* L applied to the input i to yield the system output v. That is:

$$v = L[i]$$

Linear systems exhibit some outstandingly important properties, of which the most important is termed the *superposition* principle. In the case of a DT system, this can be stated as follows.

if	input $x_1(n)$ to a linear system yields output $y_1(n)$
and	input $x_2(n)$ to the system yields output $y_2(n)$
then	input $a_1x_1(n) + a_2x_2(n)$ yields output $a_1y_1(n) + a_2y_2(n)$

or, more succinctly:

$$L[a_1x_1(n) + a_2x_2(n)] = a_1L[x_1(n)] + a_2L[x_2(n)] \tag{1.15}$$

This statement of the superposition principle can be decomposed into two parts. If $a_2 = 0$ (and a_1 and x_1 are simply replaced by a and x respectively):

$$L[ax(n)] = aL[x(n)] \tag{1.16}$$

which is the *proportionality* property. Also, if $a_1 = a_2 = 1$:

$$L[x_1(n) + x_2(n)] = L[x_1(n)] + L[x_2(n)]$$

which is the *additivity* property.

Thus, for superposition to be satisfied (and hence for a system to be linear) both proportionality and additivity properties must hold.

The implications of this principle in electronic engineering are profound. For instance, in circuit analysis we can treat large and small signals separately (often exploiting dramatically simplified equivalent circuits as a result) and finally sum the individual components. In the present context of signal processing systems, a complex input signal can (by the Fourier theorem) be considered as a summation of sine and cosine waves of appropriate phase and the output found by superposition of the individual responses. The implication is that the *frequency response* – specifying how sine or cosine waves are modified in magnitude and phase as a function of frequency – is an outstandingly useful description of a linear system. We will have much occasion to refer to the magnitude and phase responses of DT systems throughout this book.

In addition to superposition, another very important property of a linear system is *frequency preservation*. According to this property, the only frequency components present at the output of the system are those present at the input. In other words, there are no internally generated frequencies. (The reader may be familiar with so-called second-harmonic distortion which occurs in the output of audio amplifiers driven by high amplitude inputs into a region of non-linear operation.)

The reason for this property lies in the fact that a term of the form e^{st} (s complex) has the same form when differentiated with respect to time. Hence, if the forcing function in equation (1.14) is of this form, so will be the system output. Now, using the fundamentally important Euler relation $e^{j\theta} = \cos\theta + j\sin\theta$, sine and cosine waves can be represented in complex exponential form as:

$$\sin(\omega t) = \frac{1}{2j}\left(e^{j\omega t} - e^{-j\omega t}\right)$$

$$\text{and } \cos(\omega t) = \frac{1}{2}\left(e^{j\omega t} + e^{-j\omega t}\right)$$

and, therefore, decaying and growing sine and cosine waves can be similarly represented as sums and/or differences of complex exponentials. For example:

$$e^{\sigma t}\sin(\omega t) = \frac{1}{2j}\left(e^{(\sigma + j\omega)t} - e^{(\sigma - j\omega)t}\right)$$

Consequently, the form of any of these types of wave is preserved in a linear system.

The function e^{st} clearly has some special status in connection with linear systems and the differential equations which describe them. The reader who has studied linear differential equations may recognise e^{st} as the *eigenfunction* of the equation (or of the system which it describes). If the output of a system is equal to a complex constant times the input to a system, then that input is by definition the system's eigenfunction. The complex constant is the system's *eigenvalue*. Note that here by 'constant' we mean that the eigenvalue is not a function of time: we will see later that it is a function of frequency.

It has been convenient to describe the properties of superposition (proportionality and additivity) and frequency preservation with reference to CT systems. Subsequently, we will see that DT systems display the same essential properties. Just as a linear CT system can be described by a linear differential equation, so can a linear DT system be described by a so-called *difference equation* which is linear and has constant coefficients. This is the topic of the next chapter. Because of the close parallel between differential and difference equations, a linear DT system displays superposition and frequency preservation as does a CT system.

1.8.2 Time-invariance

Systems can also be classified as either *time-invariant* or time-varying. Systems which are both linear and time-invariant are called LTI. As the name clearly suggests, the properties of a time-invariant system do not alter with time. Suppose $y(n)$ is the output of an LTI system in response to input $x(n)$. If the input is delayed by k sample periods then, since the system's properties do not change in this time, the only effect is to delay the output similarly. Hence $y(n-k)$ is the output in response to input $x(n-k)$.

1.8.3 Unit-sample or 'impulse' response

From equation (1.6), the input to an LTI system can be written:

$$x(n) = \sum_{k=-\infty}^{\infty} x(k)\delta(n-k)$$

But:

$$y(n) = L[x(n)] = L\left[\sum_{k=-\infty}^{\infty} x(k)\delta(n-k)\right]$$

and, from the proportionality property of equation (1.16):

$$y(n) = \sum_{k=-\infty}^{\infty} x(k)L[\delta(n-k)]$$

Now, the output $h(n)$ of a system in response to the unit-sample $\delta(n)$ is called the *unit-sample* or *impulse response*, i.e. $h(n) = L[\delta(n)]$. The impulse response is of great importance in describing a DT system since, in view of equation (1.6) and the superposition principle, the response to *any* input $x(n)$ can be found as the sum of responses to a set of scaled Kronecker delta functions.

Now, for an LTI system by the principle of time-invariance:

$$\begin{aligned} &\text{if} \quad h(n) = L[\delta(n)] \\ &\text{then} \quad h(n-k) = L[\delta(n-k)] \end{aligned}$$

Hence:

$$y(n) = \sum_{k=-\infty}^{\infty} x(k)h(n-k) \tag{1.17}$$

This important equation is called the *convolution-sum*. From (1.17), we can (assuming zero initial conditions) obtain the output for any input. Hence, an LTI system is fully specified by its unit-sample (or 'impulse') response.

The convolution-sum states that the output $y(n)$ at sample instant n for a particular input sequence $x(n)$ can be obtained by reversing the impulse response in time (because the sign of k is negative in $h(n-k)$), registering this reversed sequence with the input sequence according to the index k, multiplying together the aligned impulses and summing the resulting products. To obtain the complete output sequence $y(n)$, the process is repeated for all k values for which there is overlap between $x(k)$ and $h(n-k)$ (but note that the overlap can be infinite.)

This may seem like a cumbersome procedure, and we will later develop some alternative ways of evaluating output sequences. However, direct convolution using equation (1.17) is occasionally worthwhile (when the degree of overlap is finite, of course).

The convolution operation is conventionally denoted $*$, so that equation (1.17) is compactly written:

$$y(n) = x(n) * h(n)$$

It is straightforward to show that the convolution operation is commutative. That is, $x(n) * h(n) = h(n) * x(n)$. This is left as an exercise for the reader (exercise 1.7).

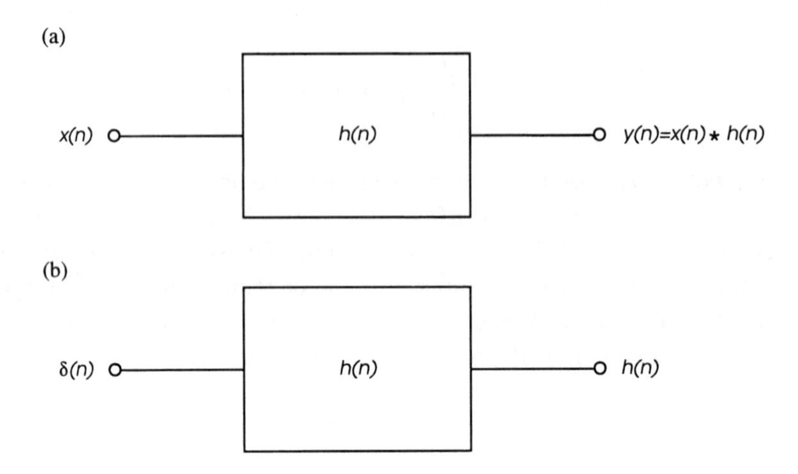

Fig. 1.8 (a) The output of a linear, time-invariant discrete-time system is found by convolving its input with its impulse response. (b) When the input is itself a unit sample, the output is the system's impulse response.

Figure 1.8(a) depicts in block diagram form how the output of a DT system is related to its input via convolution with the system's unit-sample response. By definition, the unit-sample response obtains when the input is $\delta(n)$. Figure 1.8(b) illustrates this situation. The figure also makes plain that convolution of a sequence with the unit sample is an identity operation, leaving the sequence unchanged (see exercise 1.7).

It is worth noting that convolution is an irreversible operation, in the sense that two sequences are coalesced into one. We cannot tell from the convolved sequence, what the two original sequences were. If, however, we know one of the two, then we can find the other by the inverse process of *deconvolution* which we will study later. A common situation is that we want to identify a system, i.e. find its impulse response. If we know both the input sequence and the corresponding output sequence, deconvolution will yield the impulse response. The impulse response can then be used as a *model* of the system, to find the output corresponding to any other input. In many situations, however, the input is not at all easily accessible. (A good example would be speech considered as the output from the speech production system.) In these circumstances, it may be necessary to model the system producing the signal using data from physical investigations.

1.8.4 Stability and causality

The fundamental properties of linearity and time-invariance define an important class of (LTI) discrete-time systems which can be described by the convolution-sum. From a practical point of view, there exists a most important sub-class of LTI systems which satisfy the additional properties of stability and causality.

A system is *stable* if and only if every bounded input produces a bounded output. It can be shown that this is equivalent to satisfying the condition that the impulse response is absolutely summable:

$$\sum_{n=-\infty}^{\infty} |h(n)| < \infty$$

A *causal* system is one for which the output at time index n_0 depends only upon the input for $n \leq n_0$. Consequently, an LTI system is causal if and only if its impulse response $h(n)$ is equal to zero for all $n < 0$. If this were not the case, there would be some output which *preceded* the application of the unit-sample at $n = 0$. This is not realisable in a 'real-time' system. Non-causal systems are perfectly possible, but they require 'future' values of the input to be available. That is, they work off-line, processing a history of inputs which occurred some time ago and which is stored in memory.

1.9 Fourier descriptions of CT signals

A significant proportion of this book is dedicated to the study of Fourier descriptions of discrete-time signals. To define a starting point for this, we review here Fourier descriptions of continuous-time signals. It is assumed that the reader has some familiarity with this topic through previous study.

Fourier signal representations involve the decomposition of the signal into a sum of so-called *basis functions*. In view of the earlier identification of e^{st} as the eigenfunction of a linear differential equation, complex exponential – or (co)sine – components are a natural choice for this.

1.9.1 Fourier series

Suppose we wish to represent a periodic CT function $x(t)$ in terms of complex exponential basis functions of different frequency. If $x(t)$ has period $T = 1/f_0$ (not to be confused with the sampling interval T), then its Fourier series is:

$$x(t) = \sum_{k=-\infty}^{\infty} c_k e^{2\pi jkf_0 t} \tag{1.18}$$

Clearly, equation (1.18) is periodic with period $1/f_0$ as required. To show that the (complex) coefficients c_k can be chosen so as to represent $x(t)$ exactly, we proceed as follows. Multiply both sides by $e^{-2\pi jlf_0 t}$ and integrate over a single period:

$$\int_{t_0}^{t_0+T} x(t)e^{-2\pi jlf_0 t}dt = \int_{t_0}^{t_0+T} \sum_{k=-\infty}^{\infty} c_k e^{2\pi jkf_0 t} e^{-2\pi jlf_0 t}dt$$

Interchanging the order of integration and summation on the right-hand side:

$$\int_{t_0}^{t_0+T} x(t)e^{-2\pi jlf_0 t}dt = \sum_{k=-\infty}^{\infty} c_k \int_{t_0}^{t_0+T} e^{2\pi j(k-l)f_0 t}dt$$

For $k \neq l$, the right-hand side becomes:

$$\begin{aligned}
\sum_{k=-\infty}^{\infty} c_k \left[\frac{e^{2\pi j(k-l)f_0 t}}{2\pi j(k-l)f_0}\right]_{t_0}^{t_0+T} &= \sum_{k=-\infty}^{\infty} c_k \left(\frac{e^{2\pi j(k-l)f_0(t_0+T)} - e^{2\pi j(k-l)f_0 t_0}}{2\pi j(k-l)f_0}\right) \\
&= \sum_{k=-\infty}^{\infty} c_k \left(\frac{e^{2\pi j(k-l)f_0 t_0}(e^{2\pi j(k-l)f_0 T} - 1)}{2\pi j(k-l)f_0}\right) \\
&= \sum_{k=-\infty}^{\infty} c_k \left(\frac{e^{2\pi j(k-l)f_0 t_0}(e^{2\pi j(k-l)} - 1)}{2\pi j(k-l)f_0}\right) \\
&\qquad\qquad (\text{since } f_0 T = 1) \\
&= 0 \qquad (\text{since } e^{2\pi j(k-l)} = 1)
\end{aligned}$$

However, if $k = l$:

$$\begin{aligned}
\int_{t_0}^{t_0+T} x(t)e^{-2\pi jlf_0 t}dt &= \sum_{k=-\infty}^{\infty} c_l \int_{t_0}^{t_0+T} dt \\
&= c_l [t]_{t_0}^{t_0+T} \\
&= c_l T
\end{aligned}$$

and, changing l to k:

$$c_k = \frac{1}{T}\int_{t_0}^{t_0+T} x(t)e^{-2\pi jkf_0 t}dt \tag{1.19}$$

This is the Fourier *analysis* equation, according to which $x(t)$ can be decomposed into a set of (co)sine waves whose magnitude and phase is given by the c_k coefficients, where the index k is the *harmonic number* multiplying f_0. Together with the Fourier *synthesis* equation (1.18), which specifies how to construct $x(t)$ from the set $\{c_k\}$, it forms a so-called *pair*.

If $x(t)$ is real (as well as periodic), it is apparent from equation (1.19) that c_k and c_{-k} are complex conjugates (and c_0 is real). This allows us to write the Fourier *synthesis* equation (1.18) in the well-known trigonometric forms which are right-sided (see exercise 1.8). We can also represent $x(t)$ graphically by its magnitude and phase spectra, i.e. plots of $|c_k|$ and $\angle c_k = \theta_k$ versus frequency. Because the c_k only exist for integer values of k, these are *line spectra*. That is, the spectral components are discrete and spaced by $1/T$ in frequency.

Sufficient (but not necessary) conditions for the c_k in equation (1.19) to yield the correct value of $x(t)$ in the Fourier *synthesis* equation (1.18) are the so-called *Dirichlet conditions*:

- $x(t)$ has only a finite number of discontinuities in any period;
- $x(t)$ contains a finite number of maxima and minima in any period;
- it is *absolutely integrable*:

$$\int_{t_0}^{t_0+T} |x(t)|dt < \infty$$

in any period.

If these conditions hold, the synthesis equation (1.18) converges to $x(t)$ (except where $x(t)$ is discontinuous, when we obtain the midpoint value of the discontinuity.)

The Fourier series uses complex exponentials as basis functions. More generally, let us denote the basis functions as $\phi_k(t)$. Then the synthesis equation will be:

$$x(t) = \sum_{k=-\infty}^{\infty} c_k \phi_k(t)$$

Basis functions satisfying the condition:

$$\int_{t_0}^{t_0+T} \phi_k(t)\phi_l(t)dt = \begin{cases} ||\phi_k||^2 & k = l \\ 0 & k \neq l \end{cases}$$

(where $||\phi_k||^2$ is a real value called the L_2-*norm*) are called *orthogonal* or *orthonormal*. From the foregoing, it is clear that the complex exponential basis functions have this property, and form an orthogonal set with L_2-norm:

$$||\phi_k||^2 = T$$

In the case of an orthonormal set, the coefficients of the synthesis function can be simply evaluated as:

$$c_k = \frac{1}{||\phi_k||^2} \int_{t_0}^{t_0+T} x(t)\phi_k(t)dt$$

exactly as we have found.

The description of a signal in terms of orthogonal basis functions also has implications for the calculation of the average power. Since $x(t)$ is periodic, it has infinite total energy but finite average power as given by equation (1.12). However, just because it is periodic, we can as well evaluate the average power over a single period (avoiding the necessity to average over a long observation time) as:

$$P_{av} = \frac{1}{T} \int_{t_0}^{t_0+T} x^2(t)dt$$

Although we have said that $x(t)$ is real, we can nevertheless express $x^2(t)$ more generally as $x(t)$ times its complex conjugate $\tilde{x}(t)$ obtained from the synthesis equation (1.18) as:

$$\tilde{x}(t) = \sum_{k=-\infty}^{\infty} \tilde{c}_k e^{-2\pi jkf_0t}$$

Hence:

$$\begin{aligned} P_{av} &= \frac{1}{T} \int_{t_0}^{t_0+T} x(t) \sum_{k=-\infty}^{\infty} \tilde{c}_k e^{-2\pi jkf_0t} dt \\ &= \sum_{k=-\infty}^{\infty} \tilde{c}_k \frac{1}{T} \int_{t_0}^{t_0+T} x(t) e^{-2\pi jkf_0t} dt \\ &= \sum_{k=-\infty}^{\infty} \tilde{c}_k c_k \qquad \text{from equation (1.19)} \end{aligned}$$

and:

$$P_{av} = \sum_{k=-\infty}^{\infty} |c_k|^2 \tag{1.20}$$

This is a particular form of a more general relation called *Parseval's theorem*. In this case – Parseval's relation for a finite-power signal – $|c_k|^2$ is the average power in the kth harmonic of the signal. Equation (1.20) states that the average power in the signal is obtained by summing *independently* the power in the harmonics. This independence is a direct consequence of the orthogonality of the basis functions: when we square $x(t)$ and integrate to evaluate the average power, the products of basis functions with different harmonic number integrate to zero.

Equation (1.20) suggests an alternative representation of $x(t)$ – as a plot of $|c_k|^2$ against frequency. For obvious reasons in view of the above, this is called the *power spectrum*. If $x(t)$ has units of volts then, clearly, the components of the power spectrum will have units of V^2.

1.9.2 The Fourier transform

As we have just seen, the class of (finite-power) periodic CT signals can be represented by a sum of (co)sinusoidal components of appropriate magnitude and phase, and harmonically related to the fundamental frequency. The Fourier series describes the synthesis of such a periodic CT signal from its components. Is there, then, a comparable description for the class of aperiodic, finite-energy signals? In fact, there is, and it is of paramount importance in the study of signals and systems. We call it the Fourier transform (FT).

Consider the finite duration (and, therefore, finite-energy) signal $x(t)$ depicted in figure 1.9(a). This can be thought of as the limiting case of a periodic signal $x_P(t)$ – as shown in figure 1.9(b) – as its repetition period T tends to infinity. This suggests a Fourier representation for such a signal which is likewise the limiting case of the Fourier series as the repetition period tends to infinity. This is the Fourier transform. Since the periodic signal has a line spectrum, with the harmonic components spaced by $1/T$, the aperiodic signal will have a continuous spectrum.

Now, from the analysis equation (1.19) with $t_0 = -T/2$:

$$\begin{aligned} c_k &= \frac{1}{T}\int_{-T/2}^{T/2} x(t)e^{-2\pi jkf_0t}dt \\ &= \frac{1}{T}\int_{-T/2}^{T/2} x(t)e^{-jk\omega_0t}dt \end{aligned}$$

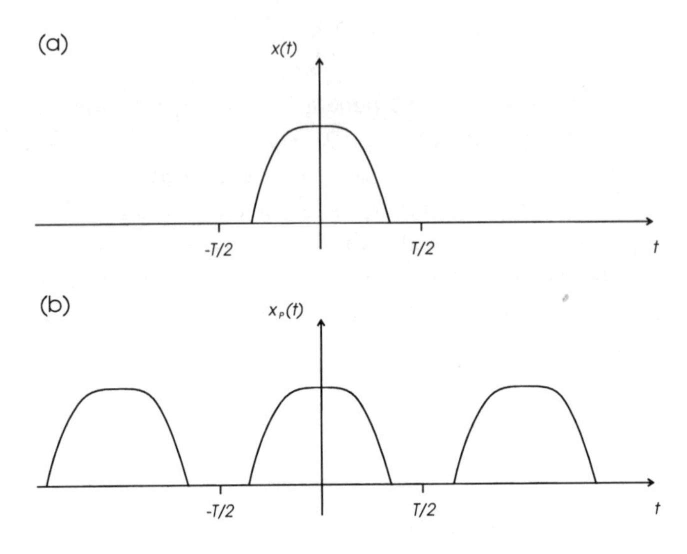

Fig. 1.9 The aperiodic signal $x(t)$ in (a) can be thought of as the limiting case of a periodic signal $x_P(t)$ of period T as T tends to infinity.

Let us write the product $c_k T$ as a new variable X. As T tends to infinity, the limits of integration tend to $\pm\infty$ but ω_0 tends to 0. Also, $k\omega_0$ tends to a continuous function of frequency, which we will call ω. Hence:

$$X(j\omega) = \int_{-\infty}^{\infty} x(t)e^{-j\omega t}dt \tag{1.21}$$

This very important equation defines the Fourier transform of $x(t)$ denoted:

$$\mathcal{F}[x(t)] = X(j\omega)$$

Consider next the synthesis equation (1.18), which becomes:

$$\begin{aligned} x(t) &= \sum_{k=-\infty}^{\infty} \frac{X(j\omega)}{T} e^{jk\omega_0 t} \\ &= \sum_{k=-\infty}^{\infty} X(j\omega) \frac{\omega_0}{2\pi} e^{jk\omega_0 t} \end{aligned}$$

Again, in the limit $T \rightarrow \infty$, $k\omega_0$ is replaced by ω, ω_0 by $d\omega$ and the summation becomes an integral, giving:

$$x(t) = \frac{1}{2\pi} \int_{-\infty}^{\infty} X(j\omega) e^{j\omega t} d\omega \tag{1.22}$$

which is the inverse Fourier transform (IFT):

$$\mathcal{F}^{-1}[X(j\omega)] = x(t)$$

Equation (1.21) allows us to analyse $x(t)$ into its *spectrum* consisting of a set of complex exponential (i.e (co)sine component) basis functions, whereas equation (1.22) enables us to reconstruct the time-domain signal from its spectrum. These two equations, therefore, form a *transform pair.*

The conditions for an aperiodic CT function to have an FT are similar to the Dirichlet conditions for a periodic function to have a Fourier series. In the latter case, however, $x(t)$ must be absolutely integrable over a period. For the aperiodic signal, since the FT is obtained from the Fourier series by letting the period T tend to infinity, $x(t)$ must be absolutely integrable over the entire range $\pm\infty$. A finite-energy signal satisfies the closely-related property of square integrability, which also means the FT is a valid description in this case.

For such a signal:

$$E = \int_{-\infty}^{\infty} x^2(t) dt$$

As before, although we have assumed $x(t)$ to be real, we can more generally express the energy as:

$$E = \int_{-\infty}^{\infty} x(t)\tilde{x}(t) dt \tag{1.23}$$

From equation (1.22):

$$\tilde{x}(t) = \frac{1}{2\pi} \int_{-\infty}^{\infty} \tilde{X}(j\omega) e^{-j\omega t} d\omega$$

and, substituting into equation (1.23):

$$\begin{aligned} E &= \int_{-\infty}^{\infty} x(t) \frac{1}{2\pi} \int_{-\infty}^{\infty} \tilde{X}(j\omega) e^{-j\omega t} d\omega dt \\ &= \frac{1}{2\pi} \int_{-\infty}^{\infty} \int_{-\infty}^{\infty} \tilde{X}(j\omega) x(t) e^{-j\omega t} d\omega dt \end{aligned}$$

Changing the order of integration:

$$
\begin{aligned}
E &= \frac{1}{2\pi}\int_{-\infty}^{\infty} \tilde{X}(j\omega) \int_{-\infty}^{\infty} x(t)e^{-j\omega t} dt d\omega \\
&= \frac{1}{2\pi}\int_{-\infty}^{\infty} \tilde{X}(j\omega) X(j\omega) d\omega
\end{aligned}
$$

and:

$$E = \int_{-\infty}^{\infty} x^2(t)dt = \frac{1}{2\pi}\int_{-\infty}^{\infty} |X(j\omega)|^2 d\omega$$

This is Parseval's relation for an aperiodic, finite-energy signal. It states how energy is related in the time and frequency domains.

The integrand $|X(j\omega)|^2$ is called the *energy density spectrum* of $x(t)$ since it describes how the (finite-)energy is distributed over frequency. Note carefully that we cannot (as we did with the line spectrum of the periodic, finite-power signal) sensibly refer to the energy (or power) at a given frequency. Because the finite-energy is distributed over all frequencies, the energy at any given frequency is necessarily zero. It only makes sense to refer to the energy within a particular band of frequencies as:

$$\Delta E = \int_{\omega'}^{\omega' + \Delta\omega} |X(j\omega)|^2 d\omega$$

Suppose as before that $x(t)$ has units of volts. Because we have to integrate over all frequencies to obtain the total energy, the units of any particular value of $|X(j\omega)|^2$ are V^2s/Hz. That is, the energy density is expressed per unit bandwidth.

1.10 THE LAPLACE TRANSFORM

The Fourier transform enables us to decompose an aperiodic, finite-energy CT function $x(t)$ into a spectrum $X(j\omega)$ extending over all frequencies. This makes it an outstandingly important tool in the study of signals and systems. Its usefulness is, however, limited by the fact that the Fourier integral of equation (1.21) does not converge for some functions, e.g. those that are not absolutely integrable over all time. This includes some functions which are of considerable interest, such as the unit-step function, $u(t)$.

In developing the FT, we took the basis functions into which $x(t)$ is decomposed to have the form $e^{j\omega t}$. However, we previously stated that the eigenfunctions of a general linear system have the form e^{st} where s is complex. Let us, therefore, put:

$$s = \sigma + j\omega \qquad (\sigma \text{ real})$$

in the defining equation (1.21), and also assume that $x(t)$ is a positive time function (i.e. $x(t) = 0$ for $t < 0$), to yield:

$$\begin{aligned} X(s) &= \int_0^\infty x(t)e^{-\sigma t}e^{j\omega t}dt \\ &= \int_0^\infty x(t)e^{-st}dt \end{aligned} \qquad (1.24)$$

which is the Laplace transform (LT) of $x(t)$:

$$\mathcal{L}[x(t)] = X(s)$$

From equation (1.24), we see that the LT of $x(t)$ corresponds to the Fourier transform of $e^{-\sigma t}x(t)$, provided that $x(t)$ is a positive time function. That is:

$$\begin{aligned} \mathcal{L}[x(t)] &= \mathcal{F}[e^{-\sigma t}x(t)] \\ \text{and } \mathcal{F}[x(t)] &= \mathcal{L}[x(t)]_{\sigma=0} = X(s)|_{\sigma=0} \end{aligned}$$

Since σ is real, the LT analyses $x(t)$ into a set of exponentially growing and decaying sine and cosine waves. This means that the LT avoids some of the convergence problems associated with the Fourier integral (1.21), since values of σ (> 0) which ensure convergence in $0 \le t < \infty$ will generally lead to divergence in $-\infty < t < 0$. That is, $e^{-\sigma t}$ acts as a convergence factor.

As an illustration, consider the Laplace transform of $x(t) = u(t)$:

$$\begin{aligned} \mathcal{L}[u(t)] &= \int_0^\infty e^{-st}dt \\ &= \left[\frac{-e^{-st}}{s}\right]_0^\infty \\ &= \frac{1}{s} \end{aligned}$$

By contrast, the FT of $u(t)$ is not defined since $u(t)$ is not absolutely integrable.

Consider also the Laplace transform of $\delta(t)$:

$$\mathcal{L}[\delta(t)] = \int_0^{\infty} \delta(t)e^{-st}dt$$

By the sifting property of the Dirac delta function:

$$\begin{aligned}\mathcal{L}[\delta(t)] &= e^{-st}\big|_{t=0} \\ &= 1\end{aligned}$$

Since $\delta(t)$ is the time derivative of $u(t)$, this illustrates an important property of the Laplace transform, namely that differentiation of $x(t)$ corresponds to multiplication of $X(s)$ by s.

To find the inverse Laplace transform (ILT), we note from above that:

$$\begin{aligned}e^{-\sigma t}x(t) &= \mathcal{F}^{-1}[X(s)] \\ &= \mathcal{F}^{-1}[X(\sigma + j\omega)]\end{aligned}$$

From the IFT, equation (1.22):

$$e^{-\sigma t}x(t) = \frac{1}{2\pi}\int_{-\infty}^{\infty} X(\sigma + j\omega)e^{j\omega t}d\omega$$

Multiplying both sides by $e^{j\sigma t}$

$$x(t) = \frac{1}{2\pi}\int_{-\infty}^{\infty} X(\sigma + j\omega)e^{j\sigma t}e^{j\omega t}d\omega$$

Since $s = \sigma + j\omega$:

$$\begin{aligned}ds &= jd\omega \\ \text{and } x(t) &= \frac{1}{2\pi j}\int_{\sigma - j\infty}^{\sigma + j\infty} X(s)e^{st}ds\end{aligned}$$

Hence, $x(t)$ can be synthesised via the ILT as an integration of complex exponentials weighted by $X(s)$. However, the ILT may be difficult to evaluate since it involves a complex line integral.

1.11 SUMMARY

This chapter has introduced the topic of discrete-time (DT) signals, and systems for processing them. A signal is very broadly defined as some intelligible sign conveying information, from which it is clear that the subject has immensely wide applicability.

Discrete-time signals are, in many (but not all) cases, obtained from an 'underlying' continuous-time (CT) signal by sampling. In mathematical terms, the signal $x(t)$ is represented by a sequence of equally-spaced 'samples' $x(n)$ denoting the signal value at specific instants of time, nT. The interval T between values is the sampling period and the reciprocal of T is the sampling frequency, f_s. Analogue-to-digital conversion is a closely-related process which yields sample values that, in addition to being defined only at discrete instants in time, are also discrete in amplitude. The process includes a sampler as a component. Replacement of continuous amplitude values by discrete values is a necessary step if the processing of a signal is to be implemented in digital hardware. There are many advantages to such implementations, but chief among these is software programmability which gives tremendous flexibility in developing solutions to all manner of signal-processing problems. In this book, however, we more or less ignore the implications of discrete amplitude representations.

Any mathematical description of the sampling process is fraught with difficulty because of the infinitesimal nature of a sample 'instant'. To help our understanding of the mathematics involved, we introduced the (CT) Dirac delta function, and its DT counterpart – the Kronecker delta. The Dirac delta function is defined only at an instant of time, t_o; its amplitude at this time is undefined. However, the function has unit area, so its behaviour under the integral sign is well-defined. In particular, it has a sifting property whereby multiplying a function $f(t)$ by $\delta(t - t_0)$ and integrating over all time yields the specific value $f(t_0)$. An alternative definition of the Dirac delta function is as the limiting case of a rectangular pulse, whose width tends to zero while the pulse's area remains constant. The Kronecker delta is conceptually simpler. As well as existing only at a discrete time, it has a defined amplitude of unity at this time. We then examined the related (CT) unit-step function and (DT) unit-step sequence.

Using these ideas, we considered two idealised formulations of the sampling process – instantaneous sampling and natural sampling. We also saw how the Kronecker delta could be very simply used to describe a DT signal, once it has been sampled.

Unfortunately, given a sampled data sequence (i.e. a DT signal), we cannot uniquely determine the CT signal that produced it. The basis of the problem is that the cos( ) function does not have a unique inverse, $\cos^{-1}()$. Given any sampled (co)sinusoid which is a component of some complex signal, its frequency could be any one of an infinite set of frequencies. This is the problem of aliasing. Its solution is always to assume that the lowest possible (positive) frequency is the correct one. In order to make this assumption valid, we have to satisfy the criterion implicit in the very important Nyquist sampling theorem. According to this theorem, if a signal is band-limited to frequency B and sampled with with $f_s > 2B$, then it can in principle be recovered exactly from its samples. This lead to the idea of anti-alias low-pass filtering, in order to band-limit a CT signal before sampling and subsequent processing. However, because anti-aliasing is performed in the CT domain using analogue filters, the band-limiting can never be guaranteed perfect and there will usually still be significant energy above the filter's cut-off frequency, f_c. Hence, a sampling rate of something like 4 or 5 times f_c is generally reasonable.

The dual process to sampling, that of recovering a CT signal $x(t)$ from its samples $x(n)$, is called interpolation. If the Nyquist criterion is satisfied, and the samples have not been amplitude quantised, $x(t)$ can in theory be recovered exactly by a process of ideal interpolation. In the practical situation, the samples are in fact amplitude quantised and an approximate reconstruction – called digital-to-analogue conversion – is used. One particularly simple form of this, which is often used in practice, is zero-order hold. Another is linear interpolation. The reconstruction can be described mathematically in terms of an interpolation function associated with each sample value. Each function is scaled by the sample value, positioned on the sample instant, and a summation taken over time. Linear interpolation uses a triangular function. We leave for later study the required function to achieve ideal interpolation.

We next considered the concepts of power and energy as applied to DT signals. Starting with CT signals, we contrasted those that had finite average power but infinite total energy with those having zero average power but finite total energy. The former are referred to as finite-power signals, a class which include periodic and some random signals. The latter form the class of finite-energy signals, which includes time-limited transients. The extension of this classification to DT signals is straightforward.

Attention then turned from signals to systems. The treatment of signals and systems together was justified by the fact that linear systems (an important class and, indeed, the only one we will consider in this book) can be uniquely described by a particular signal – the response to an impulse. As DT systems are the subject of later study in this book, and some familiarity is assumed with CT systems at the outset, we framed our discussion of linearity in CT terms initially. Linear CT systems can be described by linear differential equations with constant coefficients, and make excellent if idealised models of many real, physical systems. They satisfy the important property of superposition which is a generalisation of the proportionality and additivity properties. They also display frequency preservation which results from the fact that inputs of the form e^{st} are preserved at the output. This is because the time derivative of this function preserves its form, and the system is described by a linear differential equation. This form of input, therefore, has some sort of special status in the study of linear systems: it is called the eigenfunction of the system. Now, linear DT systems can be described by so-called (linear) difference equations (the topic of the next chapter). Because these are the DT counterpart of differential equations, essentially identical properties of superposition and frequency preservation are possessed by linear DT systems.

If the properties of a system do not change with time, then it is classified as time-invariant. A system which is time-invariant as well as linear is called LTI. Such systems can be described fully – apart from any initial conditions – by their response to a particular impulsive input. For a CT system, this input is the Dirac delta $\delta(t)$ and the response $h(t)$ is called the impulse response. For a DT system, the input is the Kronecker delta δ_n and the response $h(n)$ is called the unit-sample response. However, the term 'impulse response' is also widely used in connection with DT systems. The output for any input can be found from the convolution-sum. Because this involves the input sequence and the unit-sample response sequence $h(n)$ only, it follows that $h(n)$ constitutes a complete description of the DT system.

In general, to be of practical value, an LTI system should be stable. This means that the system's output is bounded for any bounded input. Another useful property is causality. A causal system is one whose output responses at time n_0 depends only upon inputs for time indices $n \leq n_0$. Non-causal systems can be useful, but they require a history of past inputs and so they work 'off-line' rather than in 'real-time'.

A significant proportion of this book is devoted to the Fourier description of DT signals. As a basis for this, we reviewed the Fourier description of CT signals on the understanding that the reader had some prior knowledge of this topic.

The Fourier series represents a periodic (finite-power) CT signal $x(t)$ as an infinite sum of weighted complex exponential ('basis') functions, $e^{jk\omega_0 t}$. These functions are (co)sinusoidal and, by virtue of the multiplier k, they are harmonic with respect to the period $T = 2\pi/\omega_0$. The weighting coefficients c_k are themselves complex and so can be described in polar form, leading to the idea of magnitude and phase spectra. Because of the harmonicity, these are line spectra – only taking values at integer multiples k of $1/T$. The fact that the complex exponential basis functions are orthogonal greatly simplifies the computation of the Fourier coefficients. The Fourier series itself can be thought of as a synthesis equation, specifying how $x(t)$ is to be synthesised from the (infinite) set of amplitude-weighted and phase-shifted basis functions. The so-called Dirichlet conditions on $x(t)$ are sufficient for the Fourier series of $x(t)$ to exist. (However, they are not necessary and some functions which violate them do have a Fourier series.) The expression for the c_k coefficients can be thought of as an analysis equation, specifying how $x(t)$ is decomposed into its basis functions. The analysis and synthesis equations constitute a pair; that is, they can be used reversibly to transform between the description of $x(t)$ as a time-domain waveform and as a frequency-domain spectrum. Usually, $x(t)$ is real and then c_0 is real, while c_k and c_{-k} are complex conjugates. This observation leads to the well-known trigonometric forms of the Fourier series which are right-sided. The orthogonality of the basis functions also has implications for the calculation of the average power of $x(t)$. This can either be computed in the time domain by integration or in the frequency domain by summation of $|c_k|^2$ terms, a fact known as Parseval's relation. Accordingly, a useful representation of $x(t)$ is a plot of $|c_k|^2$ versus frequency, which is called the power spectrum.

The Fourier series representation can only be used when $x(t)$ is periodic (and finite-power). However, a limited duration (finite duration) signal can be represented by a form of the Fourier series in which the period T approaches infinity. Similar (Dirichlet) conditions were sufficient (but not necessary) for this to be possible. In this case, the analysis equation is called the Fourier transform (FT). Because the spacing of the harmonic components in the Fourier series (line) spectrum is $1/T$, the Fourier transform spectrum $X(j\omega)$ is continuous. The basis functions which were previously of the form $e^{jk\omega_0 t}$ now have the form $e^{j\omega t}$ where ω is a continuous frequency variable. The synthesis equation has an integral definition (rather than being specified by as the Fourier series summation) and is called the inverse Fourier transform (IFT). The FT and IFT together form the Fourier transform pair. A Parseval relation in terms of energy holds between the time and frequency domains, with $|X(j\omega|^2$ describing how energy is continuously distributed over frequency. This is called the energy density spectrum of $x(t)$.

Unfortunately, the Fourier transform does not converge for some signals of practical interest, such as the unit step, $u(t)$. If, instead of using basis functions of the form $e^{j\omega t}$, we use instead the eigenfunctions e^{st} with $s = \sigma + j\omega$, then the $e^{\sigma t}$ term acts as a convergence factor. This yields the Laplace transform of $x(t)$. The Fourier transform can be viewed as a special case of the Laplace transform, i.e. for $\sigma = 0$. While the Laplace transform may converge even when the Fourier transform does not, the inverse Laplace transform can be difficult to evaluate since it involves a complex line integral.

The Fourier and Laplace transforms for CT signals have several important properties that we have not discussed here. Instead, they will be dealt with when we turn our attention to DT signals and systems. We will study in some detail how Fourier techniques can be applied in the discrete-time case, as well as building on our understanding of the Laplace transform to introduce the z-transform. Before doing so, however, we will look at the use of difference equations – the DT counterpart to differential equations – to describe sampled-data signals and signals.

1.12 Exercises

1.1 List some examples of signals, and systems for processing them, other than those given in this chapter.

1.2 Prove that a discrete-time sinusoid is periodic only if its frequency f_0 is a rational number.

1.3 Is the Nyquist sampling theorem applicable to an *aperiodic* signal? (That is, what might it mean to properly-sample an aperiodic signal?)

1.4 (a) Show that linear interpolation of a DT signal to reconstruct the underlying CT signal using equation (1.11) involves the use of a triangular interpolation function as depicted in figure 1.7.

(b) What is the interpolation function used in zero-order hold digital-to-analogue conversion?

1.5 List some signals which are:

(i) periodic but not finite-power

(ii) aperiodic but finite-power

Is it possible to have signals which are *neither* finite-power *nor* finite-energy?

1.6 Starting from the definition of a linear system as one that is described by an nth-order linear differential equation, deduce the superposition principle of equation (1.15).

Show that a system described by the differential equation:

$$\frac{dy}{dt} = kxy$$

is non-linear.

1.7 Show that the convolution operation is commutative, that is:

$$x(n) * y(n) = y(n) * x(n)$$

Also show that convolution of sequence $x(n)$ with the unit-sample sequence is an identity operation, leaving $x(n)$ unchanged.

1.8 Given that the complex coefficients, c_k, of the Fourier series satisfy the condition:

$$c_{-k} = \tilde{c}_k$$

obtain the trigonometric forms of the series:

$$\text{(i)} \quad x(t) = c_0 + 2\sum_{k=1}^{\infty} |c_k| \cos(2\pi k f_0 t + \theta_k)$$

$$\text{(ii)} \quad x(t) = c_0 + \sum_{k=1}^{\infty} (a_k \cos(2\pi k f_0 t) - b_k \sin(2\pi k f_0 t))$$

where θ_k in (i) is equal to $\angle c_k$. Find a_k and b_k in terms of c_k, θ_k, etc. as appropriate.

What is the practical importance of form (ii) when $x(t)$ has either even or odd symmetry about $t = 0$?

2

Difference equations

In the previous chapter, we saw how the process of sampling a continuous-time (CT) signal $x(t)$ yields a discrete-time (DT) signal $x(nT)$ which only takes a defined value at the sampling instants, with spacing determined by the sampling period T. We also saw how, in signal processing, we are concerned with the relationship between an input signal to a system and the output from that system. This relationship describes the processing properties of the system. In the DT case, the system's input and output will be sampled signals, usually denoted $x(n)$ and $y(n)$ respectively, with T implied. To describe the relationship between these time-varying quantities mathematically, we use *difference equations*, the topic of this chapter. These are the discrete-time counterparts of the linear differential equations used in the previous chapter to describe a CT system. If the DT system under study is linear and time-invariant (LTI), the difference equation describing it will be linear with constant coefficients.

To make these ideas concrete, we consider two simple but instructive examples: a moving-average filter and an infinite impulse response filter.

2.1 A MOVING-AVERAGE FILTER

A common requirement in signal processing is the reduction of noise. Figure 2.1(a) shows a 156 Hz sine wave of unit amplitude (in some arbitrary units) which has been sampled at 5 kHz. In figure 2.1(b), we see this sine wave contaminated by the addition of random noise. (We will have a good deal to say about the description of such signals in chapter 8: for the present, we merely state that the noise signal is rectangularly dis-

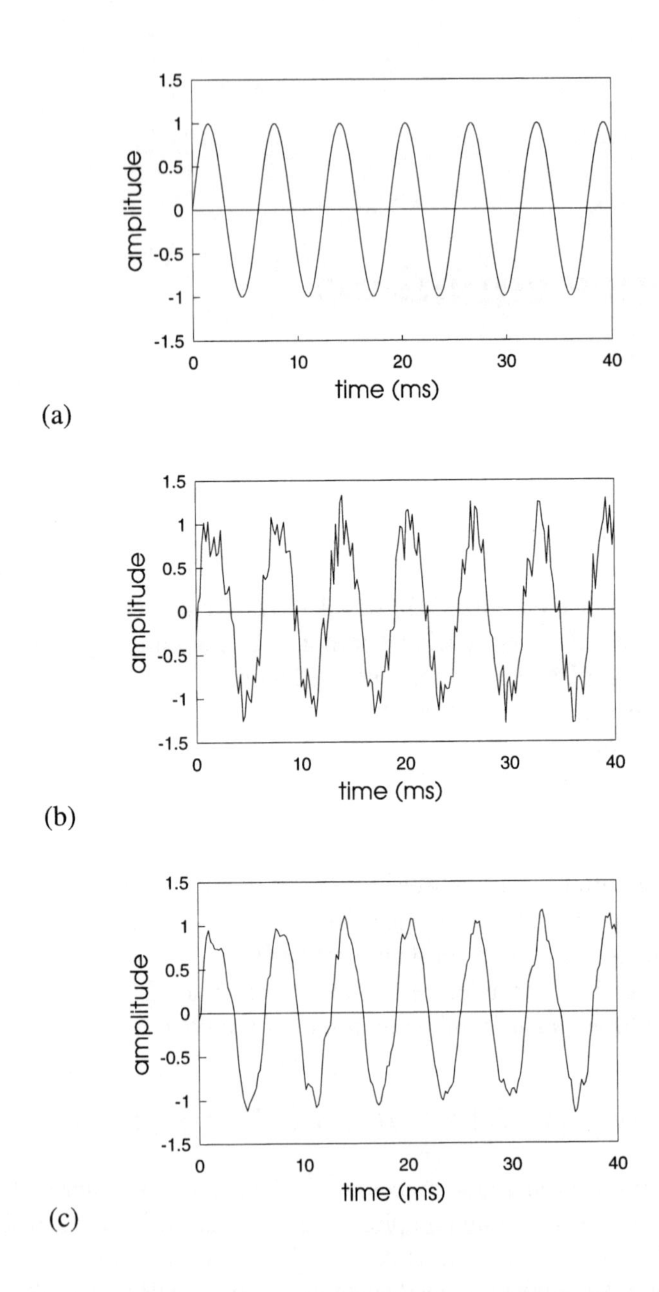

Fig. 2.1 (a) Original sampled sine wave. (b) Sine wave contaminated with random noise. (c) Effect of smoothing noisy data using a 3-point moving-average filter.

tributed in the range ±0.35.) How might we smooth, or filter, the sampled data so as to reduce the random fluctuation due to the noise and make the underlying signal more apparent?

Intuitively, it makes good sense to smooth the data by replacing any given value by the average of that value and its immediate neighbours. That is, in the form of a difference equation:

$$y(n) = \frac{1}{3}\left(x(n+1) + x(n) + x(n-1)\right)$$

The plethora of brackets in this equation make its appearance rather confusing, so here (and in general when we write a difference equation) we use our alternative subscript notation:

$$y_n = \frac{1}{3}\left(x_{n+1} + x_n + x_{n-1}\right) \qquad (2.1)$$

The key point to note about equation (2.1) is that the output is determined by the inputs only, and is independent of previous outputs.

Figure 2.1(c) shows the sequence of y_n values which results from applying the difference equation (2.1) to the noisy input data of figure 2.1(b). To do this, we have had to make assumptions about *initial* and *final* conditions. In fact, at the extreme end of the data window, where one of the neighbouring sample values is missing, the average of the two available points has simply been used. As we can see, this treatment of the data has reduced the contaminating noise very considerably, although it has not removed it entirely. Another very important point to note is that the phase of the filtered sequence is unaltered relative to that of the input. We call such a filter a *zero-phase* system.

Because of the way it is applied at each sample instant (n) in turn in calculating the average y_n, a system implementing such a difference equation is called a *moving-average filter*. In this particular case, it is a 3-point filter.

The difference equation can be considered as a prescription, or algorithm, for implementing the filter. As such, it is a very useful description of the DT system. In the previous chapter, however, we mentioned that the unit-sample (or 'impulse') response to the Kronecker delta input was also an important system description. Let us evaluate this response and see what connection there is between these two.

As described in the previous chapter, it is usual to denote the particular output to the Kronecker delta, or unit-sample sequence, δ_n (i.e. the impulse response) as h_n in place of the more general y_n. Since δ_n corresponds to an x_n which equals 1 when $n = 0$, but is zero for all other values of n (figure 2.2(a)), the output h_n will only take non-zero values for $n = -1, 0$ and $+1$. As seen in figure 2.2(b), the impulse response consists of the sequence $\{1/3,\ 1/3,\ 1/3\}$ centered around $n = 0$.

We can make a number of points about this response to the unit-sample sequence:

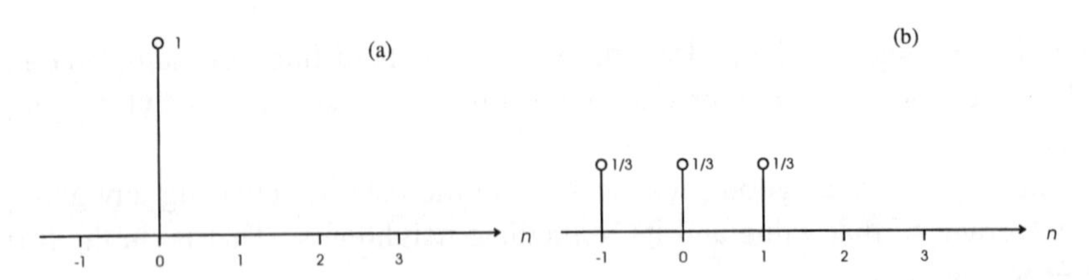

Fig. 2.2 (a) The unit-sample sequence, or Kronecker delta. (b) Unit-sample (or 'impulse') response of the 3-point moving-average filter.

- The output sequence h_n is identical to the set of coefficients in the difference equation. As we will see, this turns out to be a general and important property of systems (like this) where the current output is independent of previous outputs. (Exercise (2.5) explores this property in some detail).

- As a direct consequence of this identity, the impulse response is finite in length, having only three terms. It might seem odd at this stage to emphasise this rather obvious fact but, again, we will see that this is a key property of systems whose outputs are determined solely by their inputs. Accordingly, they are called *finite impulse response*, or FIR, systems.

- The sequence h_n sums to 1: this is an obvious consequence of our choice of the absolute values of the coefficients. Thus, the system is *stable* in the sense defined in the previous chapter because $\sum_{n=-\infty}^{\infty} |h_n| < \infty$.

- Finally, there is a non-zero response of $\frac{1}{3}$ at $n = -1$, i.e. *before* the Kronecker delta input is applied. This arises because of the x_{n+1} term in the difference equation. Thus, the system is *non-causal* in the terms defined in the previous chapter.

Since it is non-causal, this particular moving-average filter cannot be implemented in 'real-time', i.e. in a memoryless system. Provided we are prepared to wait just one sample period (until time $(n+1)$), however (and have the required memory to store it), all the information needed to calculate y_n is available. Indeed, we had no difficulty in obtaining the output sequence in figure 2.1(b) because all the input data were available to us simultaneously. Thus, a non-causal discrete-time system can be made causal by introducing a certain number of samples delay.

2.2 AN INFINITE IMPULSE RESPONSE FILTER

We now consider a slightly more complex DT system where the output sequence y_n depends upon previous outputs y_{n-k} in addition to past inputs, according to:

$$y_n = 0.1x_n + 0.4x_{n-1} + 0.5y_{n-1} \tag{2.2}$$

That is, the system has *feedback*: a previous output at time $(n-1)$, y_{n-1}, is fed back to become an input at time n. Because of the dependence on previous outputs, a difference equation like (2.2) is sometimes called *recursive*.

As before, we will evaluate the response to the unit-sample sequence, assuming the initial condition $y_{-1} = 0$. Since for this particular input, any output sample y_n is identical by definition to h_n, we have:

$$\begin{aligned}
h_0 &= 0.1x_0 + 0.4x_{-1} + 0.5y_{-1} &&= 0.1(1) + 0.4(0) + 0.5(0) &&= 0.1\\
h_1 &= 0.1x_1 + 0.4x_0 + 0.5y_0 &&= 0.1(0) + 0.4(1) + 0.5(0.1) &&= 0.45\\
h_2 &= 0.1x_2 + 0.4x_1 + 0.5y_1 &&= 0.1(0) + 0.4(0) + 0.5(0.45) &&= 0.225\\
h_3 &= 0.1x_3 + 0.4x_2 + 0.5y_2 &&= 0.1(0) + 0.4(0) + 0.5(0.225) &&= 0.1125\\
h_4 &= \ \ldots &&= \ \ldots &&= 0.05625\\
& && &&\ \text{etc.}
\end{aligned}$$

Figure 2.3 depicts this sequence. By inspection, we can write the impulse response in closed form as:

$$h_n = \begin{cases} 0.1 & n = 0 \\ 0.1(0.5)^n + 0.4(0.5)^{n-1} & n \geq 1 \end{cases} \tag{2.3}$$

By *closed form*, we mean that we can evaluate h_n directly for any n, without first having to find $h_{n-1}, h_{n-2}, \ldots, h_0$. The $n \geq 1$ qualification is necessary to exclude the zero-time response, $h_0 = 0.1$, for which the exponent in $0.4(0.5)^{n-1}$ goes negative. This is a consequence of the delayed term x_{n-1} which means that the initial impulsive excitation δ_n has an effect one sample period later. We can re-write equation (2.3), avoiding the need to include the $n = 0$ and $n \geq 1$ conditions, by using the unit-step sequence introduced in the previous chapter:

$$h_n = 0.1(0.5)^n u_n + 0.4(0.5)^{n-1} u_{n-1} \tag{2.4}$$

From equation (2.4), it is seen that h_n dies rapidly away, tending to zero exponentially as $n \to 0$. Accordingly, this is referred to as a *transient* response, since it only has

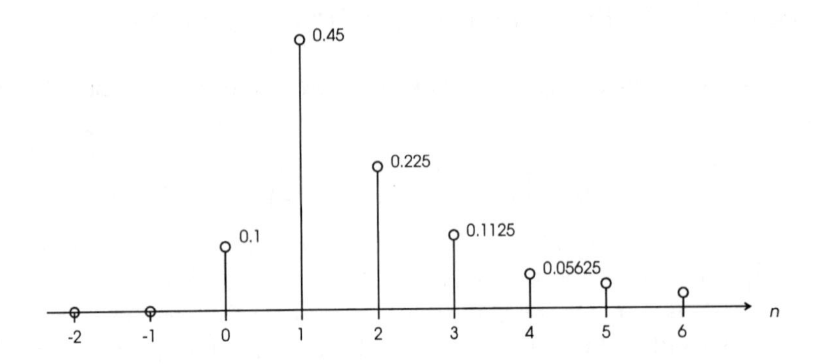

Fig. 2.3 Impulse response of discrete-time system having (recursive) difference equation $y_n = 0.1x_n + 0.4x_{n-1} + 0.5y_{n-1}$. The response extends to infinity for $n \geq 0$.

a significant effect for a limited time after application of the input. However, although it reduces with time, the sequence never actually reaches zero. This, therefore, is an *infinite impulse response* or IIR system. The infinite response results from the feedback referred to above. We can think of this in energy terms: once we have a finite output, it keeps circulating around the feedback loop dissipating energy but never disappearing entirely. The fact that the output reduces monotonically is a consequence of the y_{n-1} coefficient (equal to 0.5) in the difference equation having magnitude less than 1. If we sum the infinite sequence h_n, we find the result is unity – see exercise (2.4). The fact that this sum is finite means that the system is, by the definition in the previous chapter, stable. IIR systems obviously have the potential to be unstable, i.e. to produce an unbounded output for a bounded input, and this is a practical drawback. By contrast, FIR systems are guaranteed stable because there is a finite number of output terms, each of which is bounded if the input is. Finally, this system is causal; there is no response to the Kronecker delta input δ_n before $n = 0$. For a right-sided input sequence $x(n)$ with $n_0 = 0$:

$$x_n = \sum_{k=0}^{\infty} x_k \delta_{nk}$$

and, hence, there is no response to *any* input before $n = 0$.

Another useful description of a system is the *step response*, s_n, i.e. the output when the input is the unit step, u_n. For the IIR system under consideration, it is easily shown by direct substitution into the difference equation that:

$$s_n = \{0.1, 0.55, 0.755, 0.8875, \ldots\}$$

and so, by inspection:

$$s_n = 1 - 0.9(0.5)^n \qquad n \geq 0$$

In this case, we see that the step response tends to 1 as $n \to \infty$. That is, it has a *steady-state* value equal to 1. This is in contrast to h_n which tends to zero with increasing n. So h_n is purely transient while s_n is the sum of both a transient part $(=-0.9(0.5)^n)$ and a steady-state part (=1).

2.3 GENERAL FORM OF DIFFERENCE EQUATION

The two examples of sections 2.1 and 2.2 suggest a general formula for the difference equation of a discrete-time system, in which the output sequence is expressed in terms of the input sequence and previous outputs as:

$$y_n = \sum_{r=0}^{M} b_r x_{n-r} - \sum_{k=1}^{N} a_k y_{n-k} \tag{2.5}$$

which is the DT counterpart of the linear differential equation (with constant coefficients) which describes a linear CT system (c.f. equation (1.14) of the previous chapter).

Here, the lower limits of 0 and 1 on the summations are meant to imply that the system is causal. As we have seen, we also need information about initial conditions. Assuming these to be zero (i.e. both x_n and y_n identically zero for $n < 0$), then equation (2.5) is a complete description of the input-output behaviour of the system – as is the impulse response. In fact, the difference equation embodies the identical information to the impulse response. This will become clearer in the next chapter. Finally, the *order* of the system (or of the difference equation which describes it) is either M or N, whichever is the larger.

We stated earlier that the difference equation can be considered as a prescription for implementing a DT system. A useful way of visualising the implementation is in the form of a block diagram. For instance, figure 2.4(a) and (b) shows the block diagrams corresponding to our two previous examples. Note that we have introduced an additional delay to make the system of figure 2.4(a) causal, as discussed in section 2.1. Thus, its difference equation is now:

$$y_n = \frac{1}{3}(x_{n-2} + x_{n-1} + x_n) \tag{2.6}$$

The only processing elements required for these implementations are delay units (memory), scaling multipliers (for multiplying by a constant) and adders. Because of

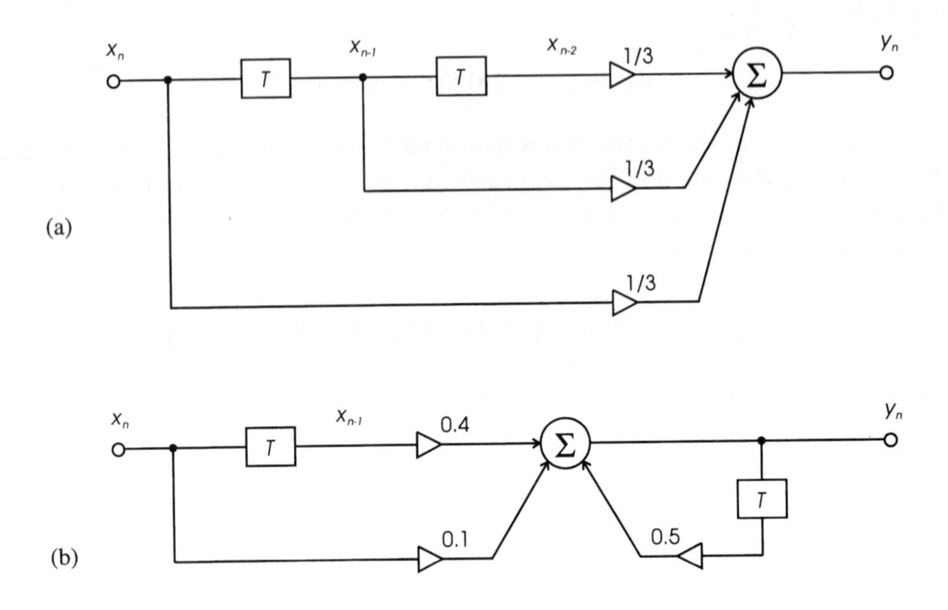

Fig. 2.4 Block diagrams of: (a) 3-point moving-average filter and (b) first-order recursive system.

the generality of equation (2.5), these are the only three elements we ever need to implement linear, time-invariant systems. The FIR structure typified by figure 2.4(a) with its absence of feedback and its cascaded delays is sometimes called a *transversal* filter because of the transverse 'taps' from the line of delay elements. By contrast, the recursive nature of the system in figure 2.4(b) is clearly seen.

While FIR filters are most naturally realised non-recursively (without feedback), there is in principle nothing to stop us devising a recursive realisation. Although a somewhat perverse thing to do, we can still achieve a finite-duration unit-sample response by ensuring that the feedback terms cancel the feedforward terms to give a zero output at some sample instant (and all future instants) when the effect of the unit sample on the feedforward terms has died away (see exercise 2.7). For this reason, some authors like to reserve the terms *recursive* and *non-recursive* for describing the implementation, rather than any fundamental property of the system.

2.4 DISCRETE DIFFERENTIATION AND INTEGRATION

Just as a linear, time-invariant CT system can be described by a differential equation with constant coefficients and no cross terms (i.e. terms like xy, $x\frac{dy}{dt}$ etc.), so a linear, time-invariant DT system can be described by a difference equation of the sort described in this chapter. Because difference equations are the DT counterpart of differential equations, we can (as we show in section 2.5 below) solve difference equations by discrete versions of essentially the same techniques that are used for solving differential equations. The formal analogy between the two can sometimes be exploited in the design of discrete-time systems, as we now show for the example of a digital differentiator.

2.4.1 Differentiation

Consider a continuous function of time, $x(t)$. According to elementary calculus, the derivative with respect to t is defined as:

$$\frac{dx}{dt} = \lim_{\Delta t \to 0} \frac{x(t + \Delta t) - x(t)}{\Delta t}$$

This suggests that we might implement a simple discrete-time differentiator by the difference equation:

$$y_n \left(\approx x'(nT)\right) = \frac{x_{n+1} - x_n}{T} \tag{2.7}$$

where the sample period T is taken as analogous to Δt. The obvious problem with this simple formulation is that the derivative is defined in terms of the limit as Δt tends to zero: hence, the closeness of the approximation depends upon the sampling frequency. In order to improve the approximation, we would have to increase the sample rate so as to decrease T. It would be much preferable to reduce the dependence of the result on T as much as possible. We can do this using the well-known Taylor series expansion.

A general CT function $f(t)$ can be expressed as a power series expanded about the point t_0:

$$f(t) = \sum_{n=0}^{\infty} a_n h^n \tag{2.8}$$

where $h = (t - t_0)$. Assuming f is differentiable to any arbitrary degree, we can differentiate (2.8) term-by-term to give:

$$
\begin{aligned}
f'(t) &= \sum_{n=1}^{\infty} n a_n h^{n-1} \\
f''(t) &= \sum_{n=2}^{\infty} n(n-1) a_n h^{n-2} \\
\text{and} \qquad f^{(k)}(t) &= \sum_{n=k}^{\infty} \frac{n!}{(n-k)!} a_n h^{(n-k)}
\end{aligned}
$$

Putting $t = t_0$ (i.e. $h = 0$):

$$
f^{(k)}(t_0) = \sum_{n=k}^{\infty} \frac{n!}{(n-k)!} a_n 0^{(n-k)}
$$

The only non-zero term is that for which the exponent of h (= 0) is 0, i.e. $n = k$, giving:

$$
\begin{aligned}
f^{(k)}(t_0) &= k! a_k \\
\text{and} \qquad a_k &= \frac{f^{(k)}(t_0)}{k!}
\end{aligned}
$$

Substituting back into the original power series, equation (2.8), gives the Taylor series expansion:

$$
f(t) = \sum_{n=0}^{\infty} \frac{f^{(n)}(t_0)\, h^n}{n!} \tag{2.9}
$$

Thus, we can approximate the continuous function $x(t)$ in the region of $t = t_0$ by truncating the Taylor series:

$$
x(t) \approx x(t_0) + (t - t_0) \frac{dx(t_0)}{dt} + \frac{(t-t_0)^2}{2!} \frac{d^2x(t_0)}{dt^2} + \cdots + \frac{(t-t_0)^N}{N!} \frac{d^N x(t_0)}{dt^N} \tag{2.10}
$$

where the degree of approximation depends upon the closeness of t to t_0 and the value of N defining the point of truncation.

Equation (2.10) is now to be recast in discrete-time form. In this case, the term $(t - t_0)$ can be considered as equivalent to n samples spaced by the sample period T, as depicted in figure 2.5. Thus, substituting $t = (t_0 + nT)$ in equation (2.10) and using the notation x' for dx/dt gives:

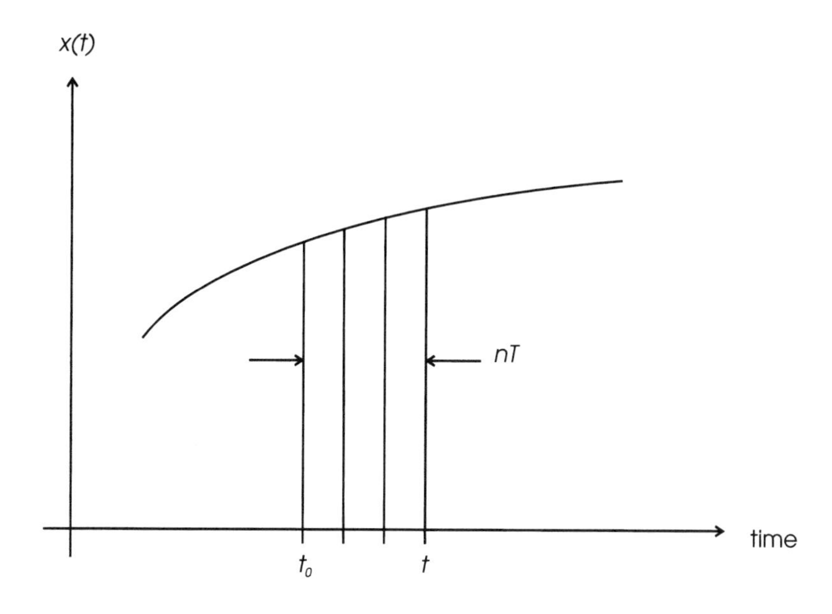

Fig. 2.5 Taylor's expansion can be used to approximate a function $x(t)$ at a point t_0 close to t. The expansion can be recast into discrete-time form by representing $(t - t_0)$ as n samples spaced by the sampling period T.

$$x\left(t_0 + nT\right) \approx x\left(t_0\right) + nTx'\left(t_0\right) + \frac{(nT)^2}{2!}x''\left(t_0\right) + \cdots \tag{2.11}$$

We can, without loss of generality, make t_0 equal to 0. Then, setting n equal to -1 gives:

$$x_{-1} = x_0 - Tx_0' + \frac{T^2}{2!}x_0'' + O\left(T^3\right) \tag{2.12}$$

where $O(T^3)$ denotes an error term "of order T^3" – signifying that, for small T, the error will depend primarily upon a term involving T^3. By comparison, terms in T^4 and higher will be small. Including this $O(T^3)$ term means we can replace the $\approx$ relation by an equality.

Likewise, with $t_0 = 0$ and setting $n = -2$ in equation (2.11):

$$x_{-2} = x_0 - 2Tx_0' + \frac{4T^2}{2!}x_0'' + O\left(T^3\right) \tag{2.13}$$

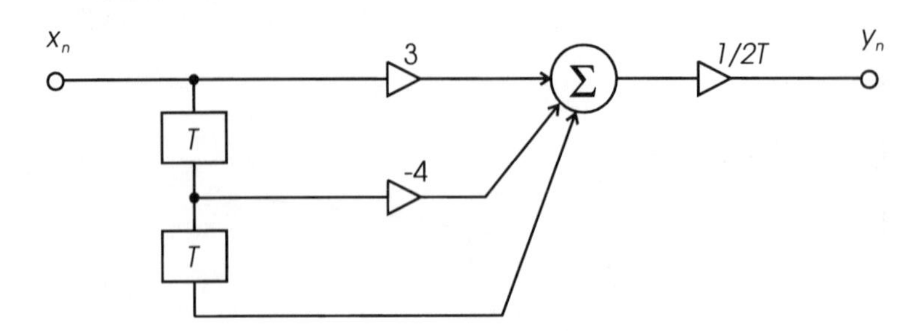

Fig. 2.6 Block diagram of discrete differentiator designed using the Taylor series expansion.

It is important to realise that the $O(T^3)$ term in equation (2.13) is *not* equal to the $O(T^3)$ in equation (2.12). The $O()$ notation merely indicates the order of magnitude of the approximation error; it does not specify an exact value.

We can now eliminate x_0'' from these two equations by subtracting 4 times equation (2.12) from equation (2.13):

$$x_{-2} - 4x_{-1} = -3x_0 + 2Tx_0' + O\left(T^3\right)$$

$$\text{or} \quad x_0' \approx \frac{3x_0 - 4x_{-1} + x_{-2}}{2T}$$

In general, by the principle of time-invariance (section 1.8.2), we have for any n:

$$y_n \left(\approx x_n'\right) = \frac{3x_n - 4x_{n-1} + x_{n-2}}{2T} \tag{2.14}$$

We expect this to be a rather better approximation to x_n' than is equation (2.7), because the error term in T^2 in the latter case is replaced by an error term in T^3.

While the error involved in using this system can be qualitatively characterised as $O(T^3)$, we can estimate it rather more quantitatively as follows. From the Taylor expansion, the error term in T^3 is actually $\frac{(nT)^3}{3!}x_0'''$. In equations (2.12) and (2.13), these evaluate to $\frac{-T^3}{3!}x_0'''$ and $\frac{-8T^3}{3!}x_0'''$ respectively. When subtracting 4 times equation (2.12) from equation (2.13), this gives a term of $\frac{-4T^3}{3!}x_0'''$ which we assume will dominate the overall error – the other terms in the expansion being of order T^4 and higher.

Figure 2.6 shows the discrete-time differentiator in block diagram form. It is an FIR system as, intuitively, the nature of differentiation tells us it must be. Whatever the conceptual difficulties of defining the derivative of $\delta(t)$ in continuous-time and/or that of δ_n in discrete-time (and they are considerable!), it is clear that it should only be non-zero close to zero time. Note also that the differentiator is causal, using only past sample values.

As a further example, let us design a discrete-time system to compute the second derivative of its input using a 3-point difference equation. In this case, we let $t_0 = 0$ as before and substitute $n = 1$ in equation (2.11). Subtracting equation (2.12):

$$\begin{aligned} x_1 + x_{-1} &= 2x_0 + \frac{2T^2}{2!}x_0'' + O\left(T^3\right) \\ \text{or} \qquad x_0'' &\approx \frac{x_1 - 2x_0 + x_{-1}}{T^2} \end{aligned}$$

More generally, the difference equation is:

$$y_n \left(\approx x_n''\right) = \frac{x_{n+1} - 2x_n + x_{n-1}}{T^2}$$

Again, this is an FIR system but it is non-causal, involving future as well as past values.

2.4.2 Integration

One useful approach to the design of a discrete-time integrator is to exploit the fact that definite integration (between two limits) of a continuous-time function evaluates the area under the curve of the function between those limits.

Considering figure 2.7, for the CT case we can make the present output of the integrator equal to the output at some previous time plus an estimate of the area under the curve of the input function between that previous time and the current instant. We attempt to do something analogous in the DT case:

$$y_n = y_{n-k} + A \tag{2.15}$$

where A denotes the 'area under the curve' of $x(nT)$ between x_{n-k} and x_n. Of course, 'area' is a problematic concept here because a DT signal x_n is not defined between samples, i.e. for non-integer n. We are really appealing to the notion of an underlying CT signal from which x_n values have been obtained by sampling, and which is recoverable from the samples by interpolation.

The simplest integrator merely makes $k = 1$ and uses a rectangular approximation to the area:

$$y_n = y_{n-1} + x_n T \tag{2.16}$$

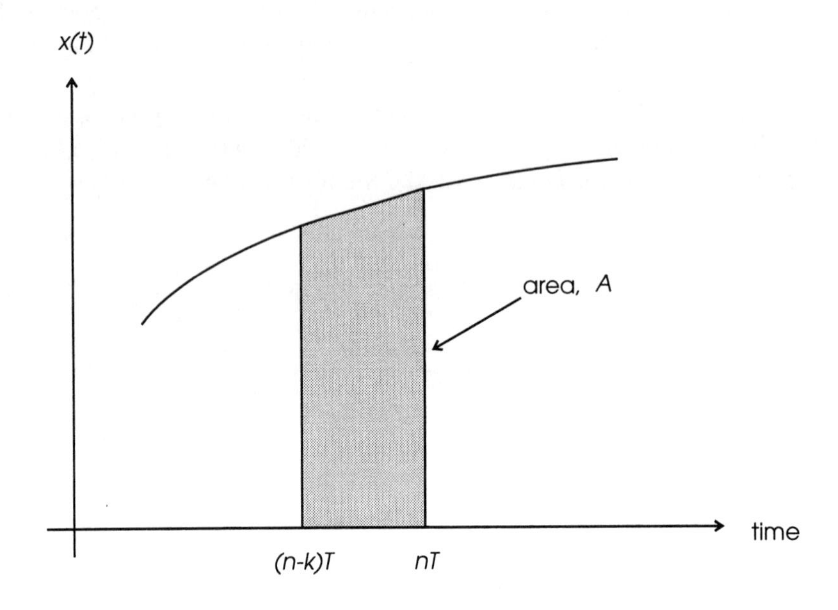

Fig. 2.7 Integration can be cast as the addition of some area measure A to a previous output value.

Note that this is identical to the zero-order hold digital-to-analogue conversion process of the previous chapter (section 1.6), emphasising the underlying importance of interpolation to discrete-time integration.

We could as well make $A = x_{n-1}T$; the effect would only be to shift the output one sample period earlier. (In the next chapter, we will use a form of rectangular approximation in which the rectangle A is centered on the sample instant.) The integrator is an IIR system; again this is a necessity given that the integral of a unit-sample at time $n = 0$ (i.e. the Kronecker delta) will ideally have a constant value of 1 for all $n \geq 0$.

A better approximation is likely to result from using a trapezoidal approximation to the area between adjacent sample points, still with $k = 1$. This can be recognised as the linear interpolation of the previous chapter. In this case, A can be broken down into a rectangular and a triangular part, to give:

$$y_n = y_{n-1} + x_{n-1}T + \frac{(x_n - x_{n-1})\,T}{2}$$

$$y_n = y_{n-1} + \frac{(x_{n-1} + x_n)\,T}{2} \tag{2.17}$$

Better still, making $k = 2$ in equation (2.15) allows a piecewise-quadratic fit to three sample values. Using the well-known Simpson's rule:

$$y_n = y_{n-2} + \frac{(x_{n-2} + 4x_{n-1} + x_n)\,T}{3} \tag{2.18}$$

It seems reasonable to expect that the performance of these integrators improves as we go from rectangular, through trapezoidal, to piecewise-quadratic approximations. Considering the rectangular approximation, this will be exact if the input x_n is a constant, i.e. if $x'_n = 0$, suggesting that the range of error will depend on the smallest and largest values of the derivative in the range of integration. By a similar argument, the error for the trapezoidal approximation will depend upon the magnitude of the second derivative, x''_n. For the piecewise-quadratic approximation, however, it turns out (the proof is beyond the scope of this book) that the error depends on $x_n^{(4)}$, i.e. the approximation is not only exact for polynomials of second order, but for those of third order also.

Finally, we note that discrete integration is often referred to as *numerical integration*. Along with discrete ('numerical') differentiation, it is a much-studied subject.

2.5 SOLUTION OF DIFFERENCE EQUATIONS

In section 2.2, we found the unit-sample (impulse) and step responses of a discrete-time system by assuming zero initial conditions and substituting input (x_n) values into the difference equation for the particular cases $x_n = \delta_n$ and $x_n = u_n$. This can be seen as solving the difference equation for y_n given the particular input and (zero) initial conditions. The type of solution which results is called *recursive*, because of the way we use previous outputs ($y_{n-1}, y_{n-2}, \ldots$) to find the current output (y_n). This is often inconvenient: we usually want to find y_n directly, i.e. in closed form, without having to find all previous outputs first. Because difference equations are the DT analogue of differential equations, we can use methods which are the DT counterparts of the classical means of solving linear differential equations with constant coefficients.

Let us try to solve equation (2.2) in this way for the same two inputs. This difference equation is first-order (c.f. equation (2.5)) since $M = N = 1$. Note also that the coefficients are constants, independent of the time indices k and r; it is this property which makes the system time-invariant. From (2.2):

$$y_n - 0.5y_{n-1} = 0.1x_n + 0.4x_{n-1} \tag{2.19}$$

We are interested here in the solutions for $x_n = u_n$ and $x_n = \delta_n$.

The classical method of undetermined coefficients solves a differential equation by finding a *complementary solution* and a *particular solution*. In system terms, the complementary solution, CS, is characteristic of the system itself and is independent of the particular input. It describes the way that energy initially stored in the system (at zero time) decays with time. In other words, it gives the transient response, since this can be viewed as the response to the zero-time conditions. The particular solution, PS, is that part of the response which is 'particular' to the specific input and has the same time dependence as it. This part of the response persists as long as the input is present. Hence, it can be equated to the steady-state response for the particular input. By the principle of superposition for linear systems, the total response is the sum of CS and PS.

We will deal with the straightforward case of the step response before turning to the rather more difficult impulse response.

2.5.1 Step response

To find the complementary solution, CS, we consider the *homogeneous equation*. This equation – because its solution depends only on the zero-time conditions and is independent of any particular input – is obtained by setting the right-hand side of equation (2.19) (the forcing function) to zero. Hence:

$$y_{CS}(n) - 0.5y_{CS}(n-1) = 0 \tag{2.20}$$

By inspection, the solution to this simple difference equation is:

$$CS = y_{CS}(n) = Af^n u_n$$

where both A and f are arbitrary constants which remain to be evaluated. The unit step u_n appears because the system is causal and, therefore, $y_n = 0$ for $n < 0$. It is common terminology to say that the signal y_n is *causal* (see exercise 2.3).

Substituting into (2.20):

$$Af^n u_n = A0.5f^{n-1}u_{n-1}$$

With any $n \geq 0$, so that both unit steps evaluate to 1:

$$f = 0.5$$

Note that the constant A disappears by cancellation; this is the sense in which the coefficient is *undetermined*. Such coefficients are not determined from the form of the complementary solution but must be found from initial conditions. So, the complementary solution is:

$$CS = A(0.5)^n u_n$$

and, hence, any energy stored initially in the system decays away exponentially.

The particular solution, PS, has the same time dependence as the input, $x_n = u_n$. So we have:

$$PS = y_{PS}(n) = Bu_n$$

Substituting into equation (2.19):

$$Bu_n - 0.5Bu_{n-1} = 0.1u_n + 0.4u_{n-1} \tag{2.21}$$

Now, since $y_n = (CS + PS)$ and CS decays to zero with increasing time:

$$\lim_{n\to\infty} y_n = y_{ss}(n) = PS$$

where $y_{ss}(n)$ is the steady-state response. Hence, equation (2.21) is evaluated with $n \to \infty$. For any $n \geq 0$, all unit steps take value 1 to give:

$$\begin{aligned} B - 0.5B &= 0.1 + 0.4 \\ \text{and} \quad B &= 1 \end{aligned}$$

The total solution is the sum of complementary and particular solutions:

$$\begin{aligned} s_n &= CS + PS \\ &= A(0.5)^n u_n + Bu_n \\ &= 1 + A(0.5)^n \quad n \geq 0 \end{aligned}$$

It only remains to find the 'undetermined' coefficient A from the initial conditions. Since, for zero initial conditions, s_n is easily evaluated as 0.1:

$$\begin{aligned} 0.1 &= 1 + A \\ \text{and } A &= -0.9 \end{aligned}$$

Hence, the total solution is:

$$s_n = 1 - 0.9(0.5)^n \qquad n \geq 0$$

as previously found.

Note that the total solution has value proportional to the particular solution once transient terms due to the system characteristics have died away with time, as $n \to \infty$. (Here, the constant of proportionality is $B = 1$.) Hence, we frequently characterise a system by its *transient* and *steady-state* responses to the unit step, u_n. Because the form of the transient response is characteristic of the system and does not change with input, it is often called the *natural response*. The steady-state response, which is determined by the input and persists as long as the input is present, is often called the *forced response*.

2.5.2 Impulse response

We turn now to the determination of the impulse response, h_n, to the input $x_n = \delta_n$. As we shall see, this is considerably harder to find but the solution does illustrate some important points.

In general, the form of the CS (equal to $A(0.5)^n u_n$) will be unchanged for a change of input. This is because the CS (i.e. the transient or natural response) is characteristic of the system and independent of the particular input. Let us make this assumption and further suppose, as earlier, that the PS has the same time dependence as the input. That is:

$$PS = B\delta_n$$

If we attempt to find B by substituting into the difference equation (2.19), we obtain:

$$B\delta_n - 0.5B\delta_{n-1} = 0.4\delta_n + 0.1\delta_{n-1}$$

which should be evaluated for $n \to \infty$. However, all terms disappear in this case. This reflects the fact that the steady-state value of the unit input – and hence of the steady-state impulse response – is zero. It seems that our solution technique is dependent upon the particular solution tending to some non-zero value in the steady-state and cannot be used here without some amendment.

Hence, we take a different tack and find A and B from known values of h_0 and h_1. We have a total solution:

$$\begin{aligned} y_n &= h_n \\ &= CS + PS \\ &= A(0.5)^n u_n + B\delta_n \end{aligned}$$

We can very easily determine (as in section 2.2) that, for the initial condition $y_{-1} = 0$, $h_0 = 0.1$. So for $n = 0$:

$$A + B = 0.1$$

Since, from earlier, $h_1 = 0.45$:

$$\begin{aligned} A(0.5) + 0 &= 0.45 \\ \text{so that} \quad A &= 0.9 \\ \text{and} \quad B &= -0.8 \end{aligned}$$

The total solution is:

$$h_n = 0.9(0.5)^n u_n - 0.8\delta_n$$

which we want to express in the form of equation (2.4) previously, i.e. without any Kronecker delta functions. Clearly:

$$h_n = 0.1(0.5)^n u_n + 0.8(0.5)^n u_n - 0.8\delta_n$$

But, from the previous chapter (equation (1.5) with $k = 1$):

$$\begin{aligned} \delta_n &= u_n - u_{n-1} \\ \text{so that} \quad u_n &= \delta_n + u_{n-1} \end{aligned}$$

and:

$$\begin{aligned} h_n &= 0.1(0.5)^n u_n + 0.8(0.5)^n (\delta_n + u_{n-1}) - 0.8\delta_n \\ &= 0.1(0.5)^n u_n + 0.8(0.5)^n \delta_n - 0.8\delta_n + 0.8(0.5)(0.5)^{n-1} u_{n-1} \end{aligned}$$

However, δ_n only takes a non-zero value for $n = 0$, so that $0.5^n \delta_n$ is simply equal to δ_n. Hence:

$$h_n = 0.1(0.5)^n u_n + 0.4(0.5)^{n-1} u_{n-1}$$

exactly as found before.

One way to interpret this expression for h_n is as a sum of two CS terms, delayed by one sample period. That is, the effect of having both x_n and x_{n-1} terms with different coefficients (0.1 and 0.5 respectively) in the difference equation (2.19) is to elicit a separate natural response for each. This perhaps gives some insight into why the method of undetermined coefficients failed to work without some amendment when applied to the unit-sample (impulse) response. The impulsive excitation to the system at times $n = 0$ and $n = 1$ is the very reason for the initial conditions which subsequently decay to zero with time. Hence, it is ambiguous as to whether the delta functions should be included with the PS (on the grounds that the forced response has the same time dependence as the input) or the CS (on the grounds that the natural response must include the initial impulsive excitation as its zero-time condition). Finally, we note that the terms 'initial' or 'zero-time' employed in conjunction with the CS are themselves slightly problematic as, in a recursive IIR system, a delta function input can effectively cause impulsive excitation at some time $n > 0$.

2.5.3 Utility of the method

The method of undetermined coefficients worked reasonably well for the case of the step response. We have only considered a first-order equation here, but the extension to second-order difference equations is straightforward. The method worked less well for the impulse response because the particular solution evaluated to zero for values of n which prevented terms disappearing to zero. In fact, there was an ambiguity as to whether the delta function input should be included in the complementary solution or the particular solution.

Although the parallel with the method for solving differential equations gives us a possible way of solving difference equations, the method is awkward and long-winded – even for inputs (like u_n) producing a non-zero steady-state output. As a consequence it is rather error-prone. It did not work without some amendment when the input was the unit-sample, or Kronecker delta, δ_n. Also, it cannot be used for difference equations of order higher than second. It happens that there is a far better method of solution based on a transform which is a close relative of the Laplace and Fourier transforms. This is the z-transform – the topic of the next chapter.

2.6 SUMMARY

In this chapter, we have seen how to describe discrete-time (DT) systems by a difference equation, which is the DT counterpart of the differential equations extensively used in the study of continuous-time (CT) systems. The difference equation is a time-domain description: it gives a sample-by-sample specification of system operation and, consequently, can be viewed as a prescription for implementation. Apart from specific initial conditions, the difference equation provides a complete description of the system's behaviour.

We applied this sample-by-sample specification to find the unit-sample (impulse) response for two simple example systems. The impulse response embodies the same information as the difference equation and so it is also a complete system description (apart from initial conditions). The first example, a 3-point moving-average filter, had a difference equation in which the only terms were input (x_n) values. It was found to describe a finite impulse response (FIR) system, and the terms of the impulse response were seen to be identical to the coefficients of the difference equation. The particular FIR system studied was non-causal, having a non-zero term in its impulse response for $n < 0$. We saw how to convert the non-causal system into a causal one by the addition of delay elements. The second example involved feedback, so that the current output became a function of previous outputs as well as of inputs. The corresponding difference equation was solved by recursion to show that the impulse response was infinite in extent. This particular infinite impulse response (IIR) system was seen to be stable because the terms of its impulse response h_n were bounded and so the $|h_n|$'s summed to a finite value. By contrast, FIR systems are unconditionally stable.

We then determined a general form for the difference equation of a linear, time-invariant DT system and showed how this related to a block-diagram description of the system. FIR and IIR systems were seen to be typified by transversal and recursive (feedback) structures respectively.

The parallel between difference and differential equations was pursued by examining some standard definitions and theorems from differential and integral calculus, and exploiting these to design DT differentiators and integrators. In particular, the theoretical basis offered by the calculus gave us a ready means of evaluating the error incurred in the designs (at least, for the differentiators).

Finally, we showed how difference equations could be solved in closed form by a DT version of the method of undetermined coefficients. This avoided any necessity to evaluate $y_{n-1}, y_{n-2}, \ldots, y_0$ when finding y_n, as required by a recursive solution. The complementary and particular solutions were equated to the transient (natural) and steady-state (forced) responses respectively. In the case of the step input to a first-order system, these solutions were evaluated without undue difficulty. Problems were found, however, when evaluating the impulse responses, essentially because the steady-state response was zero. Even without such difficulties, the method was awkward and cannot be used for difference equations of order higher than second. So (as for the convolution operation described in the previous chapter), we seek more powerful ways of obtaining solutions.

2.7 EXERCISES

2.1 By writing a computer program, or by using a spreadsheet or other suitable software, generate the sampled data depicted in figure 2.1(b). Verify that applying a 3-point moving-average filter yields the data sequence shown in figure 2.1(c). What is the effect of applying the filter twice? Experiment with the use of a 5-point moving-average filter with coefficients $\{1/5, 1/5, 1/5, 1/5, 1/5\}$. Also experiment with altering the filter coefficients such that they are unequal.

2.2 Find by recursion (if necessary) the impulse response, h_n, and the step response (i.e. the response to the input u_n), s_n, for the systems defined by each of the following difference equations:

$$\begin{aligned} \text{(i)} \quad y_n &= x_{n-4} + 4x_{n-2} + x_n \\ \text{(ii)} \quad y_n &= x_{n-1} + 0.2y_{n-2} \\ \text{(iii)} \quad y_n &= x_{n-1} + 0.2y_{n-2} + 0.2y_{n-1} \end{aligned}$$

assuming zero initial conditions.

For the two infinite impulse response (IIR) systems, express h_n in closed form.

2.3 A more formal definition of a causal (linear time-invariant) system to that given in chapter 1 is as follows. Given any two inputs $x_1(n)$ and $x_2(n)$ that are equal for all n less than some n_0, a system is causal if the corresponding outputs are also equal for $n < n_0$. Show that $h_n = 0$ for $n < 0$ is then a necessary and sufficient condition for causality. Note particularly that n_0 does not have to equal 0.

(This definition allows us to extend the concept of causality from a system to a signal. A DT signal x_n is said to be causal if $x_n = 0$ for all $n < 0$, i.e. if x_n could be the impulse response of a causal system.)

2.4 Consider the geometric series with common ratio q (which may be either real or imaginary):

$$S = \sum_{n=0}^{\infty} q^n$$

The Nth partial sum S_N, of S is given by:

$$S_N = \sum_{n=0}^{N} q^n$$

Show that:

$$S_{N-1} = \frac{1-q^N}{1-q} \qquad q \neq 1$$

and hence that, for $|q| < 1$:

$$S = \frac{1}{1-q}$$

Using this result, show that the terms of the unit-sample response of equation (2.4) sum to 1. What conditions must be satisfied by the coefficients of:

$$y_n = b_0 x_n + b_1 x_{n-1} + a_1 y_{n-1}$$

(c.f. equation (2.2)) for the corresponding DT system to be stable?

2.5 (a) For a causal finite impulse response (FIR) system, show using equation (2.5) that the terms of the impulse response, h_n, are identical to the b_r coefficients.

(b) For a stable, causal IIR system, find the particular form of equation (2.5) which describes the impulse response, h_n. Hence find the relation which holds between the a_k and b_r coefficients when $\sum_{n=0}^{\infty} h_n = 1$, and confirm that the values in equation (2.4) satisfy this relation. Why would this relation not hold if the system were unstable?

2.6 The step response of a system is given by the convolution:

$$s_n = u_n * h_n$$

Assuming a causal system with zero initial conditions, express s_n as a summation of the terms of h_n. Use this summation to check the answers obtained in exercise (2.2) above.

2.7 The most obvious and sensible way to implement an FIR system is non-recursively. This example is designed, however, to show that recursive implementations are perfectly possible (if a little contrived).

Show that the difference equation:

$$y_n = b_0 x_n + b_1 x_{n-1}$$

can be written recursively as:

$$y_n + a_1 y_{n-1} = b_0 x_n + b_2 x_{n-2}$$

and determine the values of a_1 and b_2 which make these equations equivalent.

Verify that the two equations lead to the same impulse response.

2.8 Using Taylor's theorem, show that the approximation:

$$x_n'' \approx \frac{-x_{n+2} + 16x_{n+1} - 30x_n + 16x_{n-1} - x_{n-2}}{12T^2}$$

results in an error in x'' which is two orders lower than that resulting from the alternative approximation:

$$x_n'' \approx \frac{x_{n+2} + x_{n+1} - 4x_n + x_{n-1} + x_{n-2}}{5T^2}$$

2.9 Using Taylor's theorem, form a 4-point approximation to a first derivative using the present sample value and the 3 immediate past sample values, i.e. x_n, x_{n-1}, x_{n-2} and x_{n-3}. What is the approximate error?

2.10 For the difference equations (2.16), (2.17) and (2.18), find the impulse responses of the rectangular, trapezoidal and piecewise-quadratic numerical integrators respectively. What is anomalous about the impulse response of the piecewise-quadratic integrator?

For the input $x_n = \delta_n + \delta_{n-1}$, show that the anomaly disappears. Explain this result. (Remember that the piecewise-quadratic approximation relies on fitting a curve to three adjacent sample points, i.e. over two sample intervals.)

2.11 Solve the three difference equations in exercise (2.2) by the method of undetermined coefficients, for the inputs (i) $x_n = u_n$ and (ii) $x_n = \sin(n\omega T)$. Compare the results for (i) with those found earlier by recursion.

3

The z-transform

As we saw in the previous chapter, the difference equation is a valuable way of specifying the operation of a discrete-time (DT) system on a sample-by-sample basis. However, it is very difficult from such a *time-domain* description to visualise the important, general properties of the system. In this chapter, we study a powerful way of transforming the time-domain specification using difference equations into a complex-frequency description. This latter description has many advantages: it allows us to perform convolution by simple multiplication (and so to achieve deconvolution by division), gives a powerful means of solving difference equations, and lets us visualise important properties of a DT system by plotting its poles and zeros on a complex (Argand) plane.

These are, in fact, just the sort of advantages that are gained by use of the Laplace transform (LT) in the study of continuous-time (CT) signals. While, in principle, we could also use the LT for describing discrete-time signals, there are good reasons to introduce a somewhat different transform, purpose-made for the description of DT systems and signals. This is the z-transform (ZT), which forms the topic of this chapter.

3.1 Definition

From equation (1.8) of chapter 1, we can describe the sequence produced by sampling the causal CT signal $x(t)$ with sampling period T as:

$$x^*(t) = x(t)\delta(t) + x(t)\delta(t-T) + x(t)\delta(t-2T) + \cdots \qquad (3.1)$$

We use the Dirac delta here rather than its discrete counterpart (the unit-sample sequence or Kronecker delta) because $x(t)$ is a function of the continuous time variable, t. This raises a problem with the interpretation of equation (3.1), which we previously mentioned in chapter 1. The right-hand side is a CT function yet – as a consequence of the sampling process – we would like the left-hand side to be a DT function. Indeed, $x^*(t)$ is really an abstraction, but it can be a useful one nonetheless.

Let us apply the Laplace transform to equation (3.1). The transform is defined by equation (1.24) as:

$$\mathcal{L}[x(t)] = X(s) = \int_0^\infty x(t)e^{-st}dt$$

where $s = \sigma + j\omega$, with σ acting as a convergence factor. So, for the sampled signal:

$$\begin{aligned}\mathcal{L}[x^*(t)] &= X^*(s) \\ &= \int_0^\infty [x(t)\delta(t) + x(t)\delta(t-T) + \\ &\qquad x(t)\delta(t-2T) + \cdots]e^{-st}dt \end{aligned} \tag{3.2}$$

Integrating term by term using the sifting property of the Dirac delta:

$$\begin{aligned}X^*(s) &= x(0) + x(T)e^{-sT} + x(2T)e^{-2sT} + \ldots \\ &= \sum_{n=0}^{\infty} x(nT)e^{-snT}\end{aligned} \tag{3.3}$$

Since $s = \sigma + j\omega$, the summand in equation (3.3) involves terms $e^{-j\omega nT}$ which are periodic in ω with period $2\pi/T$. Hence, $X^*(s)$ is itself periodic. This fact, as we shall see, makes use of the LT awkward for sampled-data signals and systems, especially since the periodicity depends upon T. Changing the sampling rate associated with a DT system will radically alter its description in terms of $X^*(s)$.

Thus, we seek a transformation between the s-domain and some new complex domain in which the representation of a DT signal or system is non-periodic. Let us define a new complex variable, z, in terms of $s = \sigma + j\omega$ as:

$$z = e^{sT} \tag{3.4}$$

allowing us to express the z-transform of the sampled-data signal $x(nT)$ or, more simply $x(n)$ with T implied, as:

$$\mathcal{Z}[x(n)] = X(z) = \sum_{n=0}^{\infty} x(n)z^{-n} \tag{3.5}$$

So the ZT of a sequence is an infinite power series in the complex variable z^{-1}, with the samples of the sequence appearing as coefficients. Note carefully that equation (3.5) is *not* a power series in z, since such a series strictly must have positive (and non-fractional) powers. The importance of this distinction will become clear shortly.

Since, in general, the sequence $x(n)$ will represent samples of a real signal, the coefficients of the infinite series of equation (3.5) will be real also. Furthermore, the power of z^{-1} corresponds to the sample number n of $x(n)$. Thus, each multiplication by z^{-1} can be interpreted as a shift of the sequence by one sample period. For this reason, z^{-1} is often referred to as the *shift* (or unit-delay) operator.

3.2 Taylor and Laurent series

The Taylor series expansion of a differentiable, real function is occasionally very useful, as we saw in the previous chapter where we used it in the design of discrete-time differentiators. The expansion can be very naturally extended to complex functions, when it proves to be of paramount importance in complex analysis. Accordingly, the Taylor series occupies a central place in the study of the z-transform. Before pursuing this, we review the necessary terminology from complex analysis; the reader is assumed to have some background knowledge in this subject.

A general function F of the complex variable z is *analytic* in the neighbourhood of a point z_0 in the complex plane if it has a derivative at all points in that neighbourhood. In order to have a derivative, the real and imaginary parts of z must satisfy the Cauchy-Riemann equations together with some continuity constraints. The term *domain*, which we have already used extensively if a little loosely, means all those connected points of the complex plane where $F(z)$ is analytic. The terminology 'time-domain' is a specialisation of this definition to the real axis, since time is an identically real variable. Note that z here is a *general* variable, not necessarily identical to the z (equal to e^{sT}) of the ZT although, of course, it may be. If $F(z)$ fails to be analytic at a point z_0, but is analytic within the neighbourhood of z_0, then z_0 is called a singular point, or *singularity*, of $F(z)$. For instance, $F(z) = z^{-1}$ has a singularity at $z = 0$ but is analytic elsewhere. The Dirac delta $\delta(t - t_0)$ is another example of a function which is non-analytic in its domain (the real, time axis) at $t = t_0$.

We can now extend the Taylor series of a function of a real variable (equations (2.8) and (2.9)) of the previous chapter) to the complex-variable case. If a function $F(z)$ is analytic throughout a disc-shaped region defined by $|z - z_0| < R_0$ (figure 3.1), then at

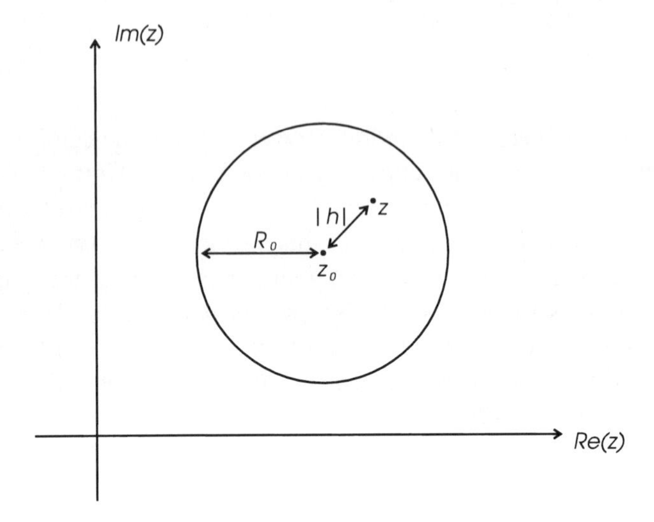

Fig. 3.1 The Taylor series for $F(z)$ converges in the disc $|z - z_0| < R_0$ when $F(z)$ is analytic at all points within that disc.

each point within that disc:

$$F(z) = \sum_{n=0}^{\infty} a_n h^n \qquad (h = z - z_0) \tag{3.6}$$

$$\text{where} \qquad a_n = \frac{F^{(n)}(z_0)}{n!} \tag{3.7}$$

exactly as for the real case.

One reason for the importance of the Taylor series in complex analysis is the fact (not proved here) that, if a complex function can be differentiated once at points in its domain D, then it can be differentiated arbitrarily many times. As a consequence, every differentiable function has a (convergent) power series expansion in D and that series is the Taylor expansion. In this case, then, convergence of the series (3.6) is assured within the region $|z - z_0| < R_0$.

One direct way to find the a_n coefficients in equation (3.7) is simply to evaluate the derivatives of $F(z)$ at the point z_0. The definition of the derivative of a complex function is a simple generalisation of the definition for the real case. Thus, the function $F(z)$ is differentiable at z_0 if the limit:

$$\lim_{z \to z_0} \frac{F(z) - F(z_0)}{z - z_0}$$

exists, in which case it is denoted $F'(z_0)$. In practice, however, it is usually much easier to evaluate the coefficients as contour integrals, as we shall see.

If $F(z)$ is continuous and has an *antiderivative* A (i.e. $A' = F$) in a domain D, then a fundamental theorem of complex analysis holds as follows. For any contour γ in D from z_1 to z_2 (the exact path from z_1 to z_2 is immaterial):

$$\int_\gamma F(z)dz = A(z_2) - A(z_1)$$

An immediate consequence is that, for a closed contour C entirely within D such that z_2 and z_1 are identical:

$$\oint_C F(z)dz = A(z_2) - A(z_1) = 0$$

A related theorem is due to Cauchy (sometimes called the Cauchy-Goursat theorem). This states that if a function $F(z)$ is analytic (rather than being continuous and having an antiderivative) at all points interior to C, then:

$$\oint_C F(z)dz = 0$$

Thus, for the contour integral to be zero, C must not enclose any singularities. (If C does enclose singularities, the integral can be computed by evaluating a constant called the *residue* of the function at each singular point, as we shall see.) For a point z_0 in D, Cauchy's integral formula states:

$$F(z_0) = \frac{1}{2\pi j} \oint_C \frac{F(z)}{(z - z_0)} dz$$

Now, by expressing $F(z)$ in the integrand as a Taylor series (equations (3.6) and (3.7)) and differentiating term-by-term, we get:

$$F^{(n)}(z_0) = \frac{n!}{2\pi j} \oint_C \frac{F(z)}{(z - z_0)^{n+1}} dz$$

and comparing with equation (3.7):

$$a_n = \frac{1}{2\pi j} \oint_C \frac{F(z)}{(z - z_0)^{n+1}} dz$$

A major limitation of the Taylor series expansion is that it cannot be used at a point z_0 if that point is a singularity. In general, this restricts us to positive powers of n (i.e. to power series). The basis of the difficulty is that while $(z - z_0)^n$ merely goes to zero at $z = z_0$ for $n > 0$, $(z-z_0)^{-n}$ goes infinite. The Taylor expansion has, however, been usefully generalised to include negative powers by Laurent. The Laurent series expansion about the point z_0 (which can be a singularity) is:

$$F(z) = \sum_{n=0}^{\infty} a_n h^n + \sum_{n=1}^{\infty} b_n h^{-n} \qquad (h = z - z_0) \tag{3.8}$$

$$\text{where} \quad a_n = \frac{1}{2\pi j} \oint_C \frac{F(z)}{(z - z_0)^{n+1}} dz$$

$$\text{and} \quad b_n = \frac{1}{2\pi j} \oint_C \frac{F(z)}{(z - z_0)^{-n+1}} dz \tag{3.9}$$

The first summation of equation (3.8) converges for $|h| < R_2$ while the second converges for $|h| > R_1$. Thus, the region of convergence for the Laurent series is:

$$R_1 < |z - z_0| < R_2$$

When $F(z)$ is actually analytic throughout the disc $|z - z_0| < R_2$, the Laurent series reduces to a Taylor series expansion about z_0.

The relevance of this to the z-transform is that the defining equation (3.5) is actually a Laurent series expansion of $F(z)$ about $z_0 = 0$, i.e. the origin. The series converges provided:

$$\sum_{n=-\infty}^{\infty} \left| f_n z^{-n} \right| < \infty$$

i.e. it is absolutely summable. From above, we know that there will be both a lower limit and an upper limit on the convergence. Representing z on the complex plane (figure 3.2), the region of convergence R is defined by:

$$R_1 < |z| < R_2$$

For our purposes, the fact that the ZT of a sequence is a Laurent series (and such series are well-studied) means that many standard results from complex analysis are directly applicable to the z-transform.

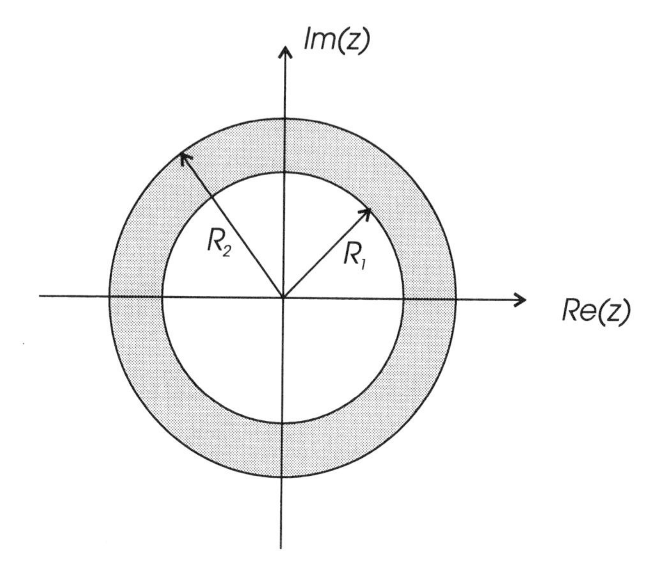

Fig. 3.2 Region of convergence of the z-transform in the complex plane.

3.3 RELATION TO THE FOURIER TRANSFORM

The relation between the z-transform and the Fourier transform (FT) of a DT sequence (which we will discuss in more detail in chapter 5) is very similar to that between the Laplace transform and the Fourier transform of a continuous function of time. However, the z-transform has been developed specifically for use with sampled-data signals and systems whereas the FT derives from the study of CT signals and systems. As we shall see, this confers considerable advantages on use of the ZT in the discrete-time case.

From chapter 1, the Fourier transform of an aperiodic CT signal, $x(t)$, is:

$$\mathcal{F}[x(t)] = X(j\omega) = \int_{-\infty}^{\infty} x(t)e^{-j\omega t}dt \tag{3.10}$$

which can be used to analyse $x(t)$ into its spectrum, consisting of a set of (co)sine components of appropriate (constant) amplitude and phase. For the transform to exist, $x(t)$ must satisfy the three Dirichlet conditions of sections 1.9.1 and 1.9.2, the third being that of absolute integrability.

As detailed in chapter 1, the essential differences between the Fourier transform of a

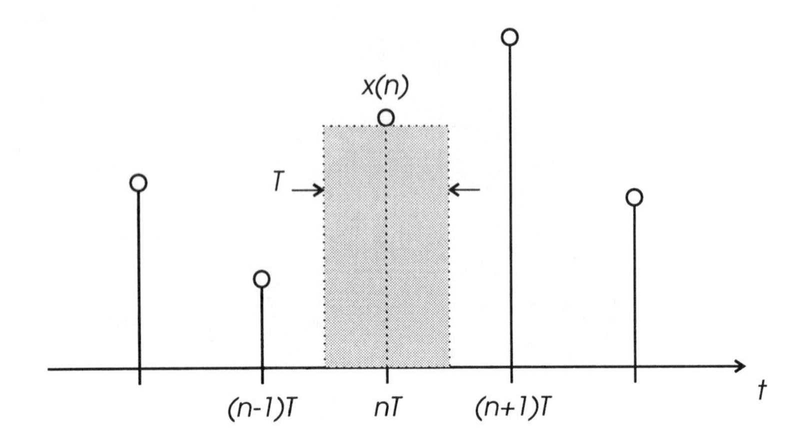

Fig. 3.3 In the case of a DT signal $x(n)$ produced from $x(t)$ by sampling, the Fourier transform can be approximated using the rectangular rule for numerical integration.

CT signal, equation (3.10), and the LT of a CT signal defined by equation (1.24) are that the integration is from $-\infty$, rather than 0, in the former and that s appears in place of $j\omega$. Thus, the LT of $x(t)$ corresponds to the FT of $e^{-\sigma t}x(t)$, provided the signal $x(t)$ is causal. Since σ is real, the LT analyses $x(t)$ into a set of exponentially growing and decaying sine and cosine waves. This means that the LT avoids some of the convergence problems associated with the FT, since values of σ (> 0) which ensure convergence in $0 \leq t < \infty$ will generally lead to divergence in $-\infty < t < 0$.

Let us now explore the corresponding relations between the z-transform and the Fourier transform of the DT signal $x(n)$ produced from $x(t)$ by sampling. To do this, we need a version of equation (3.10) appropriate for a DT signal.

One simple way to approximate the Fourier transform would be to use the rectangular integration formula of section 2.4.2 in the previous chapter. Suppose we centre rectangles of width T on the samples, $x(n)$, as depicted in figure 3.3. Thus, for a causal sequence, we can reconstruct $x(t)$ by the interpolation:

$$x(t) \approx \sum_{n=0}^{\infty} x(n)p(t - nT)$$

where the interpolation function $p(t - nT)$ is a rectangular pulse of unit height and width T, centered on the nth sample. So, equation (3.10) becomes:

$$X(j\omega) \approx \int_{-\infty}^{\infty} \sum_{n=0}^{\infty} x(n)p(t-nT)e^{-j\omega t}dt$$

Reversing the order of integration and summation, with:

$$p(t-nT) = \begin{cases} 1 & \left(nT - \frac{T}{2}\right) \le t \le \left(n + \frac{T}{2}\right) \\ 0 & \text{otherwise} \end{cases}$$

gives:

$$X(j\omega) \approx \sum_{n=0}^{\infty} x(n) \int_{nT-\frac{T}{2}}^{nT+\frac{T}{2}} e^{-j\omega t}dt$$

Now:

$$\begin{aligned}\int_{nT-\frac{T}{2}}^{nT+\frac{T}{2}} e^{-j\omega t}dt &= \left[-\left(\frac{e^{-j\omega t}}{j\omega}\right)\right]_{nT-\frac{T}{2}}^{nT+\frac{T}{2}} \\ &= -\left(\frac{e^{-j\omega nT}}{j\omega}\right)\left(e^{-j\omega T/2} - e^{j\omega T/2}\right)\end{aligned}$$

Using:

$$e^{j\omega T/2} - e^{-j\omega T/2} = 2j\sin(\omega T/2)$$

we have:

$$X(j\omega) \approx \left(\sum_{n=0}^{\infty} x(n)e^{-j\omega nT}\right) T\frac{\sin(\omega T/2)}{(\omega T/2)}$$

The $\sin(x)/x$ term is called the *sinc function*. This function occurs very commonly in discrete-time signal processing where we frequently deal with rectangular pulses and windows. We will meet it often. Note that here it is independent of n, and is not essential to the frequency-domain specification of $x(n)$. Hence, to keep the Fourier transform representation of a DT signal simple, it is usually omitted. Note too that this sinc function arises from the rectangular approximation used in the numerical integration: different approximations will yield different forms. Thus, we specify the transform as the bracketed term above:

$$X^*(j\omega) = \sum_{n=0}^{\infty} x(n)e^{-j\omega nT} \tag{3.11}$$

Comparison of this expression with equation (3.10) reveals a pleasing similarity, with the discrete variable nT replacing t and summation replacing integration. There is, however, one further important point to note about this definition of the FT. Although $X^*(j\omega)$ represents $x(n)$ as a discrete function of time (hence the $*$ superscript), it is actually a *continuous* function of frequency, ω.

Convergence of this equation requires absolute summability, i.e. that:

$$\sum_{n=-\infty}^{\infty} \left| x(n)e^{-j\omega nT} \right| \quad < \quad \infty$$

$$\text{or} \qquad \sum_{n=-\infty}^{\infty} |x(n)| \quad < \quad \infty$$

This is the discrete-time counterpart of the third Dirichlet condition of section 1.9 for the existence of the Fourier transform. The first two conditions are inapplicable in the case of a DT signal only defined at specific instants.

For a causal signal, equation (3.11) is of the same form as the defining equation for the z-transform (equation (3.5)) but with the term $e^{j\omega T}$ replacing z. This has two important implications. First, since this exponential term is periodic in ω with period $2\pi/T$, the FT (like the LT) of a sampled signal repeats indefinitely, with a period that depends upon the sampling rate. However, the ZT is non-periodic in the z-domain. It is this fundamental property which makes the ZT so suitable for the study of sampled-data signals and systems. Second, since $z = e^{(\sigma+j\omega)T}$, the ZT has a real exponential term $e^{\sigma T}$ multiplying $x(n)$ which the FT does not. This means that the ZT of a discrete-time signal may converge even when the Fourier transform does not (see example 3 of section 3.4 which follows). Of course, an exactly parallel relation holds between the Laplace transform and the Fourier transform of a CT signal.

3.4 ILLUSTRATIVE EXAMPLES

At first sight, it might appear that we have not gained much by the definition of the ZT in equation (3.5), since we still have an infinite sum to deal with. So let us explore the result of applying the transform to some typical signals. In each case, we deal with causal signals (non-zero for $n \geq 0$ only). We also include the region of convergence R since, strictly, the transform is not completely specified without it.

Example 1: Find the z-transform of the unit-sample sequence $x(n) = \delta_n$.

Substituting into the defining equation (3.5):

$$X(z) = \sum_{n=0}^{\infty} \delta_n z^{-n}$$

But the Kronecker delta, δ_{nk}, is only non-zero for $n = k$ when it is equal to 1. Here, $k = 0$ and:

$$\begin{aligned} X(z) &= z^{-n}\Big|_{n=k=0} \\ &= 1 \end{aligned}$$

The sequence is of finite length (only one value!) and obviously converges unconditionally since it is independent of z.

Example 2: Find the ZT of the shifted unit-sample sequence $x(n) = \delta_{nk}$, $k > 0$.

As before, substituting into equation (3.5):

$$\begin{aligned} X(z) &= \sum_{n=0}^{\infty} \delta_{nk} z^{-n} \\ &= z^{-n}\Big|_{n=k} \\ &= z^{-k} \end{aligned}$$

The sequence is again of length 1. It converges for all $|z| \geq 0$, i.e. everywhere. The region of convergence, R, is given by $R_1 = 0$ and $R_2 = \infty$.

Example 3: Determine the ZT of the sequence $x(n) = u_n$, where u_n is the unit-step sequence:

$$u_n = \begin{cases} 0 & n < 0 \\ 1 & n \geq 0 \end{cases}$$

Proceeding as before:

$$\begin{aligned} X(z) &= \mathcal{Z}[x_n] \\ &= \sum_{n=0}^{\infty} z^{-n} = \sum_{n=0}^{\infty} \left(z^{-1}\right)^n \end{aligned}$$

This is a geometric series with common ratio $g = z^{-1}$, which converges for $|g| < 1$ (see exercise 2.4 of the previous chapter). Thus:

$$X(z) = \left(\frac{1}{1 - z^{-1}}\right) = \frac{z}{z - 1}$$

with the region of convergence defined by $|z^{-1}| < 1$, i.e. $|z| > 1$.

The sequence $x(n) = u_n$ is not absolutely summable, so that this is an important example of a case where the ZT converges although the FT does not. (Recall that we saw an exact parallel of this situation for the CT case in section 1.10.)

Example 4: Find the ZT of the so-called exponential sequence $x(n) = a^n u_n$.

Here, we assume that a is real but may be positive or negative. For this sequence, equation (3.5) becomes:

$$\begin{aligned} X(z) &= \mathcal{Z}[x_n] \\ &= \sum_{n=0}^{\infty} a^n z^{-n} = \sum_{n=0}^{\infty} \left(az^{-1}\right)^n \end{aligned}$$

Again, this is a geometric series which converges provided the common ratio az^{-1} has modulus less than 1, to give:

$$X(z) = \frac{1}{1 - az^{-1}} = \frac{z}{z - a}$$

The region of convergence is, therefore, defined by $|z| > |a|$.

Example 5: Find the ZT of $x(n) = \sin(n\omega T)$.

Expressing the sine term in complex exponential form:

$$\begin{aligned} X(z) &= \sum_{n=0}^{\infty} \left(\frac{e^{jn\omega T} - e^{-jn\omega T}}{2j}\right) z^{-n} \\ &= \frac{1}{2j}\left(\sum_{n=0}^{\infty} \left(e^{j\omega T} z^{-1}\right)^n - \sum_{n=0}^{\infty} \left(e^{-j\omega T} z^{-1}\right)^n\right) \end{aligned} \tag{3.12}$$

By geometric series:

$$\begin{aligned}
X(z) &= \frac{1}{2j}\left(\frac{1}{1-e^{j\omega T}z^{-1}} - \frac{1}{1-e^{-j\omega T}z^{-1}}\right) \\
&= \frac{z}{2j}\left(\frac{e^{j\omega T} - e^{-j\omega T}}{\left(z-e^{j\omega T}\right)\left(z-e^{-j\omega T}\right)}\right) \\
&= \frac{z \sin \omega T}{z^2 - 2z\cos\omega T + 1} \\
&\qquad \left(\text{since } \cos\omega T = \left(e^{j\omega T} + e^{-j\omega T}\right)/2\right)
\end{aligned}$$

From the summations in equation (3.12), the region of convergence R is given by $|z| > |e^{j\omega T}|$ and $|z| > |e^{-j\omega T}|$, i.e. $|z| > 1$.

In all these examples, then, we have managed to produce a closed-form expression for the transformed sequence.

3.5 FUNDAMENTAL PROPERTIES

The z-transform has many properties which are important in the study of DT systems. Some of these properties are described below.

3.5.1 Linearity

Consider a linear combination of sequences, $x_1(n)$ and $x_2(n)$. Then:

$$\mathcal{Z}\left[ax_1(n) + bx_2(n)\right] = aX_1(z) + bX_2(z)$$

This, of course, is exactly the superposition principle (equation (1.15) for a linear system.

Proof

From the defining equation (3.5):

$$\mathcal{Z}\left[ax_1(n) + bx_2(n)\right] = \sum_{n=0}^{\infty}\left(ax_1(n) + bx_2(n)\right)z^{-n}$$

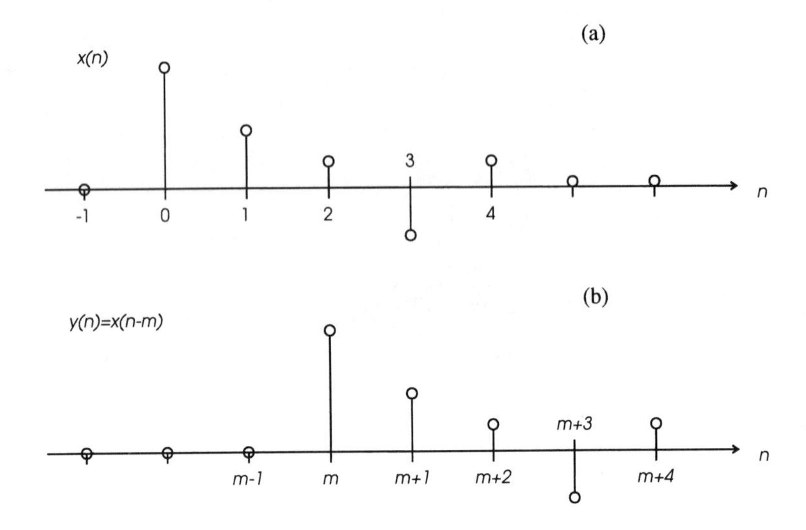

Fig. 3.4 Sequence $y(n)$ in (b) produced by delaying $x(n)$ in (a) by m sample periods.

$$= a\sum_{n=0}^{\infty} x_1(n)z^{-n} + b\sum_{n=0}^{\infty} x_2(n)z^{-n}$$

$$= aX_1(z) + bX_2(z)$$

Apart from any singularities, the region of convergence is the overlap of the individual regions of convergence of $X_1(z)$ and $X_2(z)$.

3.5.2 Right-shift property

Suppose a sequence $x(n)$ is delayed by m sample periods to give sequence $y(n)$ (figure 3.4):

$$y(n) = x(n-m) \qquad m \geq 0$$

We wish to relate the ZT of the shifted sequence, $Y(z)$, to that of the original sequence, $X(z)$. Now:

$$Y(z) = \sum_{n=0}^{\infty} x(n-m)z^{-n}$$

$$= x(-m) + x(1-m)z^{-1} + x(2-m)z^{-2} + \cdots + x(0)z^{-m} + x(1)z^{-(m+1)} + \cdots$$

But for a causal signal, $x(r) = 0$ for all $r < 0$ and:

$$\begin{aligned} Y(z) &= x(0)z^{-m} + x(1)z^{-(m+1)} + \cdots \\ &= z^{-m}\left(x(0) + x(1)z^{-1} + x(2)z^{-2} + \cdots\right) \\ &= z^{-m}\sum_{n=0}^{\infty} x(n)z^{-n} \end{aligned}$$

so that
$$Y(z) = z^{-m}X(z) \tag{3.13}$$

This is the *right-shift property*. Equation (3.13) shows that shifting a sequence rightwards by m sample periods in the time domain corresponds to multiplication of the ZT of the sequence by z^{-m}. We have, of course, already mentioned this property in section 3.1, where we interpreted z^{-1} in the ZT defining equation (3.5) as a unit-delay (or shift) operator. Also, examples 1 and 2 of section 3.4 above illustrate this property for the particular case of the Kronecker delta.

Since the actual sequence values are unaffected by any shift, the regions of convergence of $X(z)$ and $Y(z)$ are identical.

3.5.3 Convolution property

The ZT of the convolution of two sequences is equal to the product of their z-transforms. That is:

$$\begin{aligned} \text{if} \quad w(n) &= x(n) * y(n) \\ \text{then} \quad W(z) &= X(z).Y(z) \end{aligned}$$

Proof

Using the convolution-sum, equation (1.17):

$$\begin{aligned} w(n) &= x(n) * y(n) \\ &= \sum_{k=-\infty}^{\infty} x(k)y(n-k) \end{aligned}$$

$$W(z) = \sum_{n=0}^{\infty} \sum_{k=-\infty}^{\infty} x(k)y(n-k)z^{-n}$$

Reversing the order of summation:

$$W(z) = \sum_{k=-\infty}^{\infty} x(k) \sum_{n=0}^{\infty} y(n-k)z^{-n}$$

By the right-shift property above:

$$\sum_{n=0}^{\infty} y(n-k)z^{-n} = z^{-k}Y(z)$$

Hence:

$$\begin{aligned} W(z) &= \sum_{k=0}^{\infty} x(k)z^{-k}Y(z) \\ &= X(z).Y(z) \end{aligned}$$

As a corollary, we can perform deconvolution by division.

3.6 THE SYSTEM FUNCTION $H(z)$

In the previous chapter, we saw that the general difference equation for a DT system with input sequence $x(n)$ and output sequence $y(n)$ is (equation (2.5)):

$$y(n) = \sum_{r=0}^{M} b_r x(n-r) - \sum_{k=1}^{N} a_k y(n-k)$$

or

$$\sum_{k=0}^{N} a_k y(n-k) = \sum_{r=0}^{M} b_r x(n-r) \qquad \text{(with } a_0 = 1\text{)}$$

Taking z-transforms:

$$\sum_{k=0}^{N} a_k \mathcal{Z}[y_{n-k}] = \sum_{r=0}^{M} b_r \mathcal{Z}[x_{n-r}]$$

By the right-shift property:

$$\mathcal{Z}[y(n-k)] = z^{-k}Y(z)$$
$$\mathcal{Z}[x(n-r)] = z^{-r}X(z)$$

Thus:

$$\sum_{k=0}^{N} a_k z^{-k} Y(z) = \sum_{r=0}^{M} b_r z^{-r} X(z)$$

Since at least a_0 (=1) from the set of a_k coefficients is non-zero, we can divide to give:

$$Y(z) = X(z)\frac{\sum_{r=0}^{M} b_r z^{-r}}{\sum_{k=0}^{N} a_k z^{-k}}$$

But the system output is given by the convolution of the input sequence and impulse response:

$$y(n) = x(n) * h(n)$$

Taking z-transforms using the convolution property:

$$Y(z) = X(z).H(z)$$

Hence:

$$H(z) = \frac{\sum_{r=0}^{M} b_r z^{-r}}{\sum_{k=0}^{N} a_k z^{-k}} \tag{3.14}$$

$H(z)$, the z-transform of the impulse response $h(n)$, is called the *system function*. Just like the impulse response, it is a complete descriptor of a DT system (with zero initial conditions). Accordingly, we will have a good deal more to say about it. It is also known as the *transfer function*, although we will not use the term in this book. Because of the direct relation of the system function to the difference equation, the order of the system is either M or N, whichever is the larger, in accordance with the definition of the previous chapter. $H(z)$ must converge for the corresponding system to be stable.

Until now, we have applied the z-transform to sequences representing a discrete-time *signal*. The system function shows us that the ZT can be equally-well applied to the description of a DT *system* in a form of complex-frequency domain. In this case, the system function is a rational polynomial (i.e. a ratio of polynomials) in z^{-1}.

3.7 THE z-PLANE

From equation (3.14), we can express $H(z)$ as:

$$\begin{aligned} H(z) &= \frac{b_0 + b_1 z^{-1} + \cdots + b_M z^{-M}}{a_0 + a_1 z^{-1} + \cdots + a_N z^{-N}} \\ &= \frac{b_0 z^{-M}\left(z^M + \frac{b_1}{b_0} z^{M-1} + \cdots + \frac{b_M}{b_0}\right)}{a_0 z^{-N}\left(z^N + \frac{a_1}{a_0} z^{N-1} + \cdots + \frac{a_N}{a_0}\right)} \\ &= \frac{b_0 z^{-M}(z - z_1)(z - z_2)\ldots(z - z_M)}{a_0 z^{-N}(z - p_1)(z - p_2)\ldots(z - p_N)} \end{aligned} \tag{3.15}$$

The finite values of z which make $H(z) = 0$ are called the *zeros* of the system function. From equation (3.15), these include the M roots of the numerator polynomial, $z_i = \{z_1, z_2, \ldots, z_M\}$. In addition, the z^{-N} term in the denominator gives rise to N coincident zeros at the z-plane origin. We refer to this an Nth-order zero.

Likewise, the finite values of z which make $H(z)$ infinite are called the *poles* of the system function. The set of poles include the N roots of the denominator polynomial denoted $p_j = \{p_1, p_2, \ldots, p_N\}$. Again, there are also M coincident poles at the origin arising from the term z^{-M} in the numerator of equation (3.15).

The reason for the finite-value qualification in the above is that an infinite value for a zero or a pole will trivially render $H(z)$ infinite or zero, respectively.

Clearly, any function of the complex variable z which goes infinite at a point fails to be analytic at that point. Hence, the system poles are necessarily singularities of the system function $H(z)$.

Because of the factorisation in equation (3.15), and the fact that the a_k and b_r coefficients are always chosen to be real, it is apparent that poles and zeros must either be real or occur in complex-conjugate pairs.

In summary, there are $(N + M)$ finite poles (N corresponding to the $(z - p_j)$ terms and M at the origin) which are balanced by $(M + N)$ finite zeros. If the number of poles and zeros is not balanced, this has implications for the time response of the system as we shall see in section 3.10 below.

We have already found it useful to depict the region of convergence of the ZT of a sequence on a complex plane – the z-plane (figure 3.2). It can also be very instructive to plot the poles and zeros of the ZT of a sequence on the z-plane – something we will do frequently throughout the remainder of the book. We mark the zeros as circles and the poles as crosses. When the sequence represents the impulse response of a DT system, the z-plane pole-zero plot neatly depicts a wealth of information concerning the properties of the system. To see this, let us consider the relation between the z-plane and the s-plane used to describe CT signals and systems.

The s-plane to z-plane mapping, $s \to z$, is given by equation (3.4) as:

$$s \to z = e^{sT} = e^{\sigma T} e^{j\omega T}$$

Lines of constant frequency, ω, in the s-plane will, therefore, map to lines of constant angle ωT in the z-plane passing through the origin, as shown in figure 3.5 for $\omega = \omega_0$. When $\omega = 0$, z is purely real and positive so that the σ-axis of s maps to the *positive* real axis in z (figure 3.6). When $\omega = \pi/T$, z is also purely real but negative so that this particular line of constant ω maps to the *negative* real axis in z.

Now:

$$\begin{aligned} z &= e^{\sigma T}(\cos \omega T + j \sin \omega T) \\ \text{and hence } |z| &= \left|e^{\sigma T}\right| \sqrt{\cos^2 \omega T + \sin^2 \omega T} \\ &= e^{\sigma T} \qquad (\sigma \text{ is real}) \end{aligned}$$

Thus, a locus of constant σ in the s-plane corresponds to a circle of radius $e^{\sigma T}$ in the z-plane, as shown in figure 3.5 for $\sigma = \sigma_0$. In particular, the s-plane imaginary axis defined by $\sigma = 0$ maps to the unit circle in the z-plane defined by $|z| = 1$ (figure 3.6). However, the s-plane imaginary axis corresponds to the locus of real-frequency points. Since $z = e^{j\omega T}$ for $\sigma = 0$ (section 3.1), the zero frequency point corresponds to $z = 1$ in the z-plane. As ω increases, tending to ∞, we traverse the unit circle with period $\omega = 2\pi/T$. Thus, the point $z = -1$ corresponds to the infinite set of angular frequencies $\{\pi/T, 3\pi/T, 5\pi/T, \ldots\}$. Since the Laplace transform of a sampled-data sequence is periodic in ω with period $2\pi/T$, the poles and zeros of the sequence will repeat infinitely as shown in figure 3.7 for a typical case. However, as depicted, repeating poles or zeros in the s-plane each map to a single point in the z-plane, precisely because any point in the z-plane corresponds to an infinite number of frequencies spaced by $2\pi/T$. Hence, the ZT avoids the problem of infinitely repeating poles and zeros inherent in the

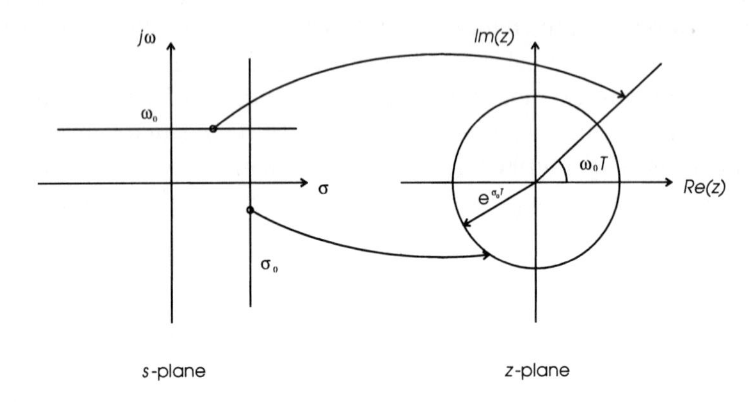

Fig. 3.5 s-plane to z-plane mapping. Lines of constant ω in s map to lines of constant angle ωT in z, whereas lines of constant σ map to circles of radius $e^{\sigma T}$ centred on the origin.

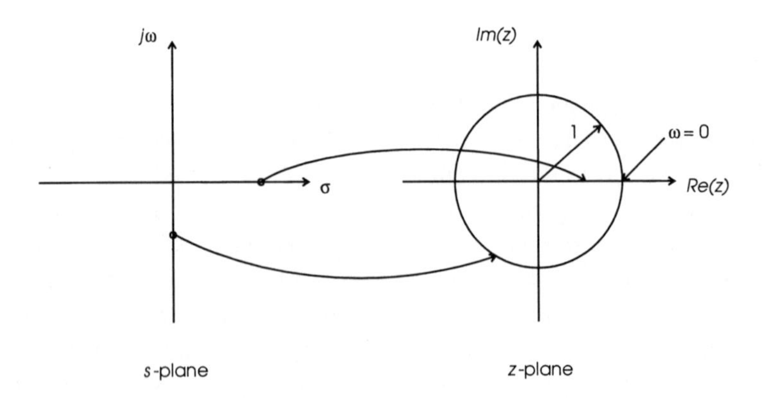

Fig. 3.6 The σ-axis of the s-plane maps to the positive real axis of the z-plane. The $j\omega$ (real frequency) axis of s maps to the unit circle of z.

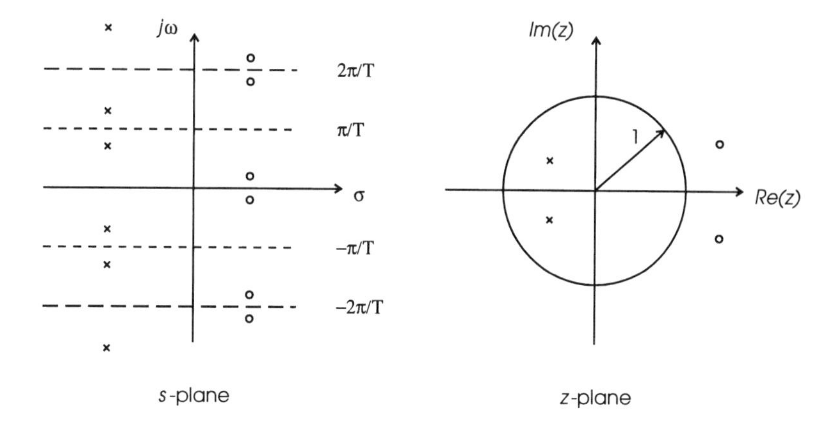

Fig. 3.7 The infinitely repetitive poles and zeros of a sampled-data sequence in the s-plane map to unique poles and zeros in the z-plane.

use of LT, by effectively overlaying them at distinct points. So, the z-domain description of a DT signal or system involves only a finite number of poles and zeros whose positions are independent of the sampling period, T.

For causal sequences, we know that the Fourier transform as in equation (3.11) corresponds to the z-transform (equation (3.5)) with z replacing $e^{j\omega T}$, i.e. with $\sigma = 0$. That is, the FT of a sequence can be found by evaluating the ZT on the unit circle. It follows that $H(z)$ must converge on the unit circle for the corresponding system to be stable, i.e. the region of convergence, R, of $H(z)$ must include the unit circle for stability to hold.

The left-half s-plane is defined by $\sigma < 0$ and so maps to the inside of the unit circle. As just stated, for stability the region of convergence of $H(z)$ must include the unit circle. In section 3.10, we show that a stable system must have its z-plane poles *within* the unit circle, paralleling the well-known result that a CT system must have its s-plane poles in the left-half plane. The right-half s-plane defined by $\sigma > 0$ maps to the outside of the unit circle. A DT system with its z-plane poles in this region will be unstable. There are no such conditions on the z-plane zeros for stability to hold.

3.8 FREQUENCY RESPONSE OF A DT SYSTEM

The frequency and time responses of a discrete-time system are of paramount interest. We postpone our treatment of the time response until we have dealt with the inverse z-transform. In this section, we study the frequency response of a DT system, which is the steady-state response (i.e. the output once initial transient responses have died down) to sinusoidal inputs. The response to such inputs is of fundamental concern since, by the Fourier theorem, any input can be constructed from sine and cosine components added with appropriate phase. This is a consequence of the fact that the eigenfunctions of a linear system are complex exponentials, as discussed in chapter 1 for the case of CT systems. We show here how the frequency response can be found from the system function $H(z)$.

3.8.1 Frequency response and the FT

Consider a DT system with impulse response $h(n)$ and whose input is the sequence $x(n)$. By the convolution-sum, the output sequence $y(n)$ is given by:

$$y(n) = \sum_{k=-\infty}^{\infty} h(k)x(n-k)$$

For a causal system, $h(k) = 0$ for $k < 0$, so that:

$$y(n) = \sum_{k=0}^{\infty} h(k)x(n-k)$$

Since we are concerned with sinusoidal inputs, let us consider the case where the input sequence is the complex exponential $x(n) = Ae^{j\omega nT}$ (A is an arbitrary real constant). Then:

$$\begin{aligned} y(n) &= \sum_{k=0}^{\infty} h(k)Ae^{j\omega(n-k)T} \\ &= Ae^{j\omega nT}\sum_{k=0}^{\infty} h(k)e^{-j\omega kT} \end{aligned}$$

which confirms that $Ae^{j\omega nT}$ is an eigenfunction of a general linear DT system. From chapter 1, the complex constant multiplying the $Ae^{j\omega nT}$ eigenfunction input is the system's eigenvalue.

By definition of the z-transform:

$$H(z) = \sum_{k=0}^{\infty} h(k)z^{-k}$$

Hence:

$$\begin{aligned} y(n) &= Ae^{j\omega nT}\, H(z)|_{z=e^{j\omega T}} \\ &= x(n)\left|H(e^{j\omega T})\right| e^{j\angle H(e^{j\omega T})} \end{aligned}$$

Thus, the magnitude and phase of the frequency response are found by evaluating the modulus and argument respectively of the system function $H(z)$ with $z = e^{j\omega T}$. This corresponds to evaluation around the unit circle, so that the frequency response is formally identical to the Fourier transform of the sequence h_n. Note the important point that although $H(z)$ is a non-periodic description of a system in the z-domain, the evaluation around the unit circle (i.e. $H(e^{j\omega T})$) is periodic. It is now apparent that the system's eigenvalue is $H(e^{j\omega T})$.

3.8.2 Magnitude and phase responses

The term $|H(e^{j\omega T})|$ is the *magnitude response* of the system and $\angle H(e^{j\omega T})$ is the *phase response*, denoted $M(\omega T)$ and $\Phi(\omega T)$ respectively.

As an example, we evaluate the frequency response of the (causal) 3-point moving-average FIR filter studied in the previous chapter whose difference equation (2.6) is:

$$y_n = \frac{1}{3}\left(x_{n-2} + x_{n-1} + x_n\right)$$

Taking z-transforms using the right-shift property:

$$Y(z) = \frac{1}{3}\left(z^{-2}X(z) + z^{-1}X(z) + X(z)\right)$$

so that:

$$\begin{aligned} H(z) = \frac{Y(z)}{X(z)} &= \frac{1}{3}\left(z^{-2} + z^{-1} + 1\right) \\ &= \frac{1}{3}z^{-2}\left(1 + z + z^2\right) \end{aligned}$$

Thus, $H(z)$ has two z-plane zeros at $z = \frac{-1\pm\sqrt{3}j}{2}$ (i.e. on the unit circle) as well as a second-order pole at the origin, as depicted in figure 3.8. In fact, because the factors $(z - p_j)$ in the denominator of equation (3.15) – which give rise to non-zero poles, p_j – are associated with feedback terms having coefficients a_k $(k \neq 0)$, an FIR system

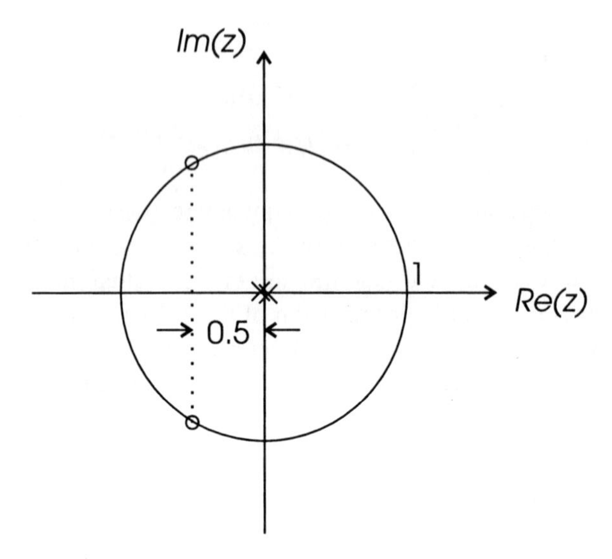

Fig. 3.8 The causal 3-point moving-average filter $y_n = \frac{1}{3}(x_{n-2} + x_{n-1} + x_n)$ has two z-plane zeros on the unit circle, and a second-order pole at the origin. Note the depiction of the second-order pole as a pair of staggered crosses.

cannot have such factors in its system function. The only poles it can have are those of the form z^{-M} in the numerator, which are located at the origin, i.e. an FIR system has no non-zero poles.

To find the frequency response, we replace z by $e^{j\omega T}$ to give:

$$\begin{aligned} H\left(e^{j\omega T}\right) &= \frac{1}{3}\left(e^{-2j\omega T} + e^{-j\omega T} + 1\right) \\ &= \frac{1}{3}(\cos 2\omega T - j\sin 2\omega T + \cos\omega T - j\sin\omega T + 1) \end{aligned}$$

We can now evaluate the magnitude response as:

$$\begin{aligned} M(\omega T) = \left|H(e^{j\omega T}\right| &= \frac{1}{3}\sqrt{Re(H)^2 + Im(H)^2} \\ \text{where} \qquad Re(H) &= \cos 2\omega T + \cos\omega T + 1 \\ Im(H) &= -\sin 2\omega T - \sin\omega T \end{aligned} \tag{3.16}$$

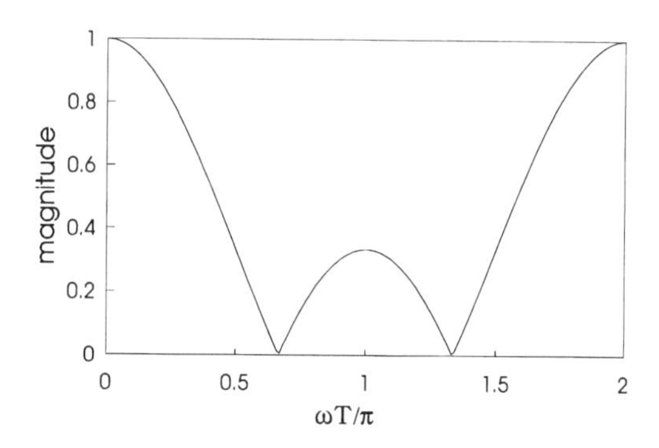

(a)

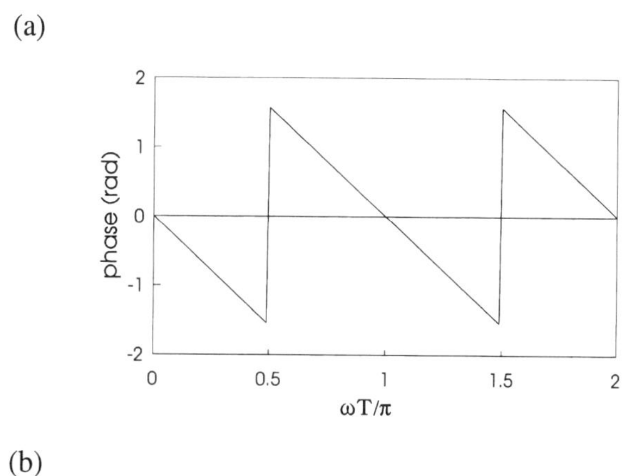

(b)

Fig. 3.9 (a) Magnitude response of 3-point moving-average filter. (b) Phase response of this filter, evaluated as $\tan^{-1}(\frac{Im(H)}{Re(H)})$.

Over the range $0 \leq \omega T \leq 2\pi$, the magnitude response is as shown in figure 3.9(a). Two important points can be made about this response. First, it is symmetrical about the point $\omega T = \pi$ corresponding to the folding (Nyquist) frequency $\omega_s/2$, and the symmetry is even:

$$M(\omega T) = M(2\pi - \omega T)$$

or equivalently:

$$M(\pi + \omega T) = M(\pi - \omega T)$$

Since the frequency response is the FT of $h(n)$ (i.e. $H(z)$ evaluated around the unit circle), and the FT of a DT signal is periodic in ω with period $2\pi/T$, we might well expect some such symmetry relation to hold.

Second, the magnitude response of figure 3.9(a) has a basically low-pass characteristic, consistent with our earlier use of the system to reduce wideband noise. Unfortunately, it is not a very impressive low-pass filter. Ideally, such a filter would have a constant transmission at frequencies up to the cut-off, with a sharp transition to zero-transmission for frequencies higher than the cut-off. Although the transmission attains a value of zero at $\omega T = \frac{2\pi}{3}$, the fall-off is rather gradual and there is a very large *side-lobe* such that the transmission of 1/3 at the folding frequency is a significant proportion of the zero-frequency gain of unity. However, the filter is exceedingly simple, and its specification was arrived at intuitively rather than on the basis of any rigorous design procedure. In later chapters, we will explore several techniques for the design of digital filters which will yield better results.

For the moment, however, let us examine the phase response of this moving-average filter. This is found from:

$$\Phi(\omega T) = \angle H(e^{j\omega T}) = \tan^{-1}\left(\frac{Im(H)}{Re(H)}\right) \tag{3.17}$$

where the real and imaginary parts are as in equation (3.16).

Figure 3.9(b) shows the phase response, again over the range $0 \leq \omega T \leq 2\pi$, evaluated using equation (3.17). The response is linear, reducing with frequency, apart from the very obvious step discontinuities at $\omega T = \pi/2$ and $3\pi/2$. Linear phase turns out to be a very desirable characteristic of a filter, as we will see in chapter 6. The discontinuities result directly from use of the inverse tangent function in equation (3.17). The $\tan^{-1}$ function is many-to-one: it returns the so-called *principal value* (an angle in the range $\pm\pi/2$) of its argument. Hence, whenever $\Phi(\omega T)$ passes through $\pm n\pi/2$ radians, a discontinuity is introduced.

Frequently in a signal-processing system it is necessary to know the actual value of phase at a given frequency rather than the principal value. (An example might be where we wish to interpolate values of the phase close to a value of $\pm n\pi/2$.) In this case, we need to employ a *phase-unwrapping* algorithm which attempts to infer the correct phase – unbounded by $\pm\pi/2$.

3.8.3 Graphical derivation of the frequency response

We have found that the frequency response of a DT system is formally equivalent to the FT (equation (3.11)) of its impulse response sequence. That is, it can be evaluated as $H(e^{j\omega T})$ where $e^{j\omega T}$ is a point on the z-plane unit circle, i.e. a vector of unit modulus and argument equal to ωT. This leads to a useful graphical means of finding the magnitude and phase responses of a system from its z-plane pole-zero diagram.

From equation (3.15):

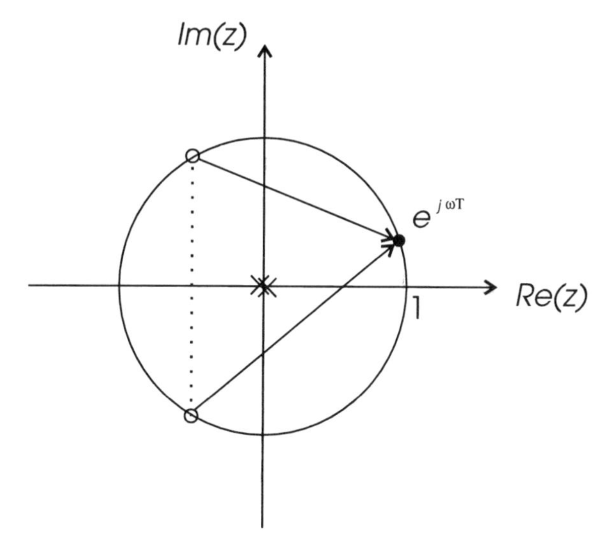

Fig. 3.10 The magnitude response at a particular frequency ω can be found graphically from vectors drawn from the (non-zero) zeros and poles of $H(z)$ to the point $e^{j\omega T}$ on the unit circle. Zeros and poles at the origin do not affect the magnitude response since the associated vectors have constant, unit modulus at all frequencies.

$$H(z) = \frac{b_0 z^{-M} \prod_{i=1}^{M} (z - z_i)}{a_0 z^{-N} \prod_{j=1}^{N} (z - p_j)}$$

so that:

$$H\left(e^{j\omega T}\right) = \frac{b_0 e^{-j\omega(M-N)T} \prod_{i=1}^{M} \left(e^{j\omega T} - z_i\right)}{a_0 \prod_{j=1}^{N} \left(e^{j\omega T} - p_j\right)}$$

where the $(e^{j\omega T} - z_i)$ and $(e^{j\omega T} - p_j)$ factors represent vectors drawn from the zeros and poles respectively to a general point $e^{j\omega T}$ on the z-plane unit circle. This is depicted in figure 3.10 for the pole-zero diagram of the example 3-point moving-average filter. Note carefully the orientation of these vectors, i.e. in the direction of the $e^{j\omega T}$ point. This turns out to be important when we consider the phase response later.

First, however, consider the magnitude response:

$$M(\omega T) = \left|H\left(e^{j\omega T}\right)\right| = \left|\frac{b_0}{a_0}\right| \left(\frac{\Pi_{i=1}^{M}\left|e^{j\omega T} - z_i\right|}{\Pi_{j=1}^{N}\left|e^{j\omega T} - p_j\right|}\right)$$

Here, $|e^{j\omega T} - z_i|$ and $|e^{j\omega T} - p_j|$ are the moduli of the vectors referred to above. Note that the zeros and poles at the origin have no effect on the magnitude response since they are identically at unit distance from any point on the unit circle.

For our illustrative 3-point moving-average filter, the magnitude response at zero frequency can be found by drawing vectors from the two zeros at $\frac{-1\pm\sqrt{3}j}{2}$ to the point $z = 1$. From simple geometry, we find that these two vectors are each of length $\sqrt{3}$. The factor $|\frac{b_0}{a_0}|$ is equal to 1/3 so that:

$$M(0) = \frac{1}{3}\sqrt{3}\sqrt{3} = 1$$

as in figure 3.9(a).

Thus, starting at the zero frequency point ($\omega T = 0$, corresponding to $z = 1$), the magnitude response can be found by evaluating the appropriate vectors as the $e^{j\omega T}$ point moves around the unit circle. This gives a very powerful way of visualising the response. For instance, it is readily apparent in the case of the moving-average filter that its magnitude response will be zero when the $e^{j\omega T}$ point is coincident with one of the zeros. This first occurs when $\omega T = 2\pi/3$, but also at $4\pi/3$, as seen in figure 3.9(a). As we will shortly show, a stable system cannot have poles on the unit circle: they must be inside it. Suppose, however, there are poles close to the unit circle. As the $e^{j\omega T}$ point approaches the vicinity of such a pole, $|e^{j\omega T} - p_j|$ will reduce to reach a minimum when this vector is radial. Since the modulus of this vector appears on the denominator, the response will peak at the corresponding frequency. The closer the pole to the unit circle, the larger this peak will be.

From this geometric interpretation, it is clear that the even symmetry of the magnitude response about the folding frequency is intimately related to the fact that poles and zeros are either real or occur in complex-conjugate pairs.

Although the phase response is not so easy to visualise in this way, we will consider it here both for completeness and because some important points emerge. The phase response is:

$$\begin{aligned}\angle H\left(e^{j\omega T}\right) &= \Phi(\omega T) \\ &= \omega(N - M)T + \sum_{i=1}^{M} \angle\left(e^{j\omega T} - z_i\right) - \sum_{j=1}^{N} \angle\left(e^{j\omega T} - p_j\right)\end{aligned}$$

Here (unlike the magnitude response), the zeros and poles at the origin do have an effect, as the appearance of the term $\omega(N-M)T$ shows. The terms $\angle(e^{j\omega T}-z_i)$ and $\angle(e^{j\omega T}-p_j)$ denote angles between the positive real axis and the vectors to the $e^{j\omega T}$ point from the (non-zero) zeros and poles respectively.

For our example 3-point moving-average filter, $N=0$ (there are no non-zero poles) and $M=2$. Hence:

$$\Phi(\omega T) = -2\omega T + \sum_{i=1}^{2} \angle\left(e^{j\omega T} - z_i\right)$$

Figure 3.11 shows four illustrative cases of $\omega T = 0, \pi/2, 2\pi/3$ and π. Let the angles of the vectors associated with the two zeros be θ_1 and θ_2 and that associated with the second-order pole be ϕ, as depicted. Then the phase response is:

$$\Phi(\omega T) = \theta_1 + \theta_2 - 2\phi$$

where ϕ is identical to ωT.

At zero frequency, it is clear from figure 3.11 that $\theta_1 = -\theta_2$ and $\phi = 0$. Hence, adding all the angles:

$$\Phi(0) = 0$$

as seen in figure 3.9(b) earlier.

At $\omega T = \pi/2$, some simple geometry gives $\theta_1 = 15^0 = \pi/12$ rad and $\theta_2 = 75^0 = 5\pi/12$ rad, so that:

$$\Phi(\pi/2) = \pi/12 + 5\pi/12 - 2(\pi/2) = -\pi/2$$

again, as seen in figure 3.9(b). As ωT increases beyond this point, $\Phi(\omega T)$ reduces below $\pi/2$. Because the $\tan^{-1}$ function returns the principal value, figure 3.9(b) displays a step discontinuity at $\omega T = \pi/2$ such that $\Phi(\omega T)$ appears to increase to a positive value which it should not do: this is actually an artifact.

Consider next the case of $\omega T = 2\pi/3$, when the $e^{j\omega T}$ point is coincident with the zero at $\frac{-1+\sqrt{3}}{2}$. At this exact frequency, the vector between the zero and the $e^{j\omega T}$ point obviously has zero length. Slightly below this frequency, the vector makes a *positive* angle, θ_1, with the real frequency axis while, slightly above it, the angle is *negative*. Hence, there is a $-\pi$ phase reversal as we pass through the zero. Let us denote the phase response at frequencies just below and just above $\omega T = 2\pi/3$ as $\Phi_-(\omega T)$ and $\Phi_+(\omega T)$ respectively. Then:

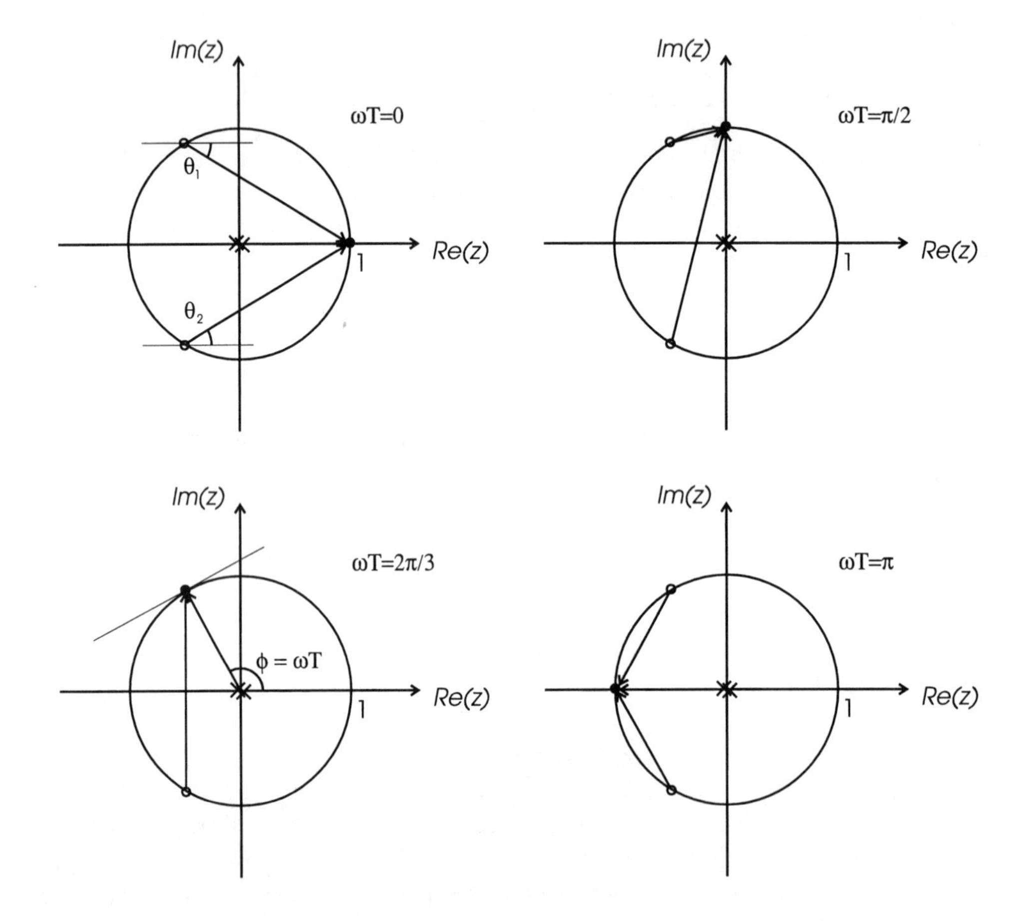

Fig. 3.11 Graphical evaluation of the phase response of the 3-point moving-average filter at particular frequencies ω.

$$\begin{aligned}
\Phi_-(\omega T) &= \Theta + \theta_2 - 2\,(2\pi/3) \\
\text{and} \quad \Phi_+(\omega T) &= (-\pi + \Theta) + \theta_2 - 2\,(2\pi/3)
\end{aligned}$$

where Θ is the angle between the positive real axis and the tangent to the unit circle at this point, as depicted in figure 3.11. It is easily seen from the figure that $\theta_2 = \pi/2$. From simple geometry, $\Theta = 30^0 = \pi/6$ rad and:

$$\begin{aligned}
\Phi_-(\omega T) &= \pi/6 + \pi/2 - 2\,(2\pi/3) \\
&= 2\pi/3 - 2\,(2\pi/3) \\
&= -2\pi/3 \\
\text{and} \quad \Phi_+(\omega T) &= (-\pi + \pi/6) + \pi/2 - 2\,(2\pi/3) \\
&= -5\pi/3
\end{aligned}$$

These two angles are not the same: they differ by $-\pi$ radians. However, their principal values are identical (equal to 60^0 or $\pi/3$ rad) so that, because of the use of the $\tan^{-1}$ function, no discontinuity is apparent in figure 3.9(b).

Of course, a similar $-\pi$ phase reversal occurs at the other zero at $\omega = \frac{-1-\sqrt{3}}{2}$.

As we will see, there is also a phase discontinuity at $\omega T = \pi$. Slightly below this frequency, the vectors from the poles at the origin make an angle of π with the positive real axis. Further, the angles θ_1 and θ_2 cancel. Hence, using our earlier subscript notation:

$$\Phi_-(\pi) = -2\pi$$

However, as ωT increases very slightly above π, this angle itself undergoes a sign reversal to become $-\pi$. Thus:

$$\Phi_+(\pi) = -\,(-2\pi)$$

and there is a $+4\pi$ phase discontinuity.

Putting all this together, the overall phase response is as shown in figure 3.12. Note that the response has odd symmetry:

$$\begin{aligned}
\Phi(\omega T) &= -\Phi(2\pi - \omega T) \\
\text{or} \quad \Phi(\pi + \omega T) &= -\Phi(\pi - \omega T)
\end{aligned}$$

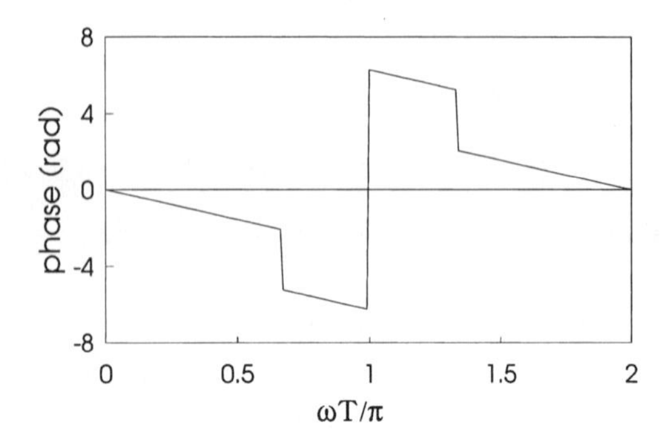

Fig. 3.12 Phase response of the 3-point moving-average filter evaluated graphically. Note the $-\pi$ phase discontinuities caused by the zeros on the unit circle. Also, the phase response has odd symmetry over the range $0 \leq \omega T < 2\pi$ rad.

in contrast to the even symmetry of the magnitude response. Again, the symmetry is due to the zeros and poles being either real or occurring in complex-conjugate pairs. The odd nature of the symmetry arises because of the phase reversal of the angle $\phi = \omega T$ as it passes through π.

We have now dealt with the frequency response of a linear, time-invariant DT system. Before going on to consider the time response, we must first study the inverse z-transform.

3.9 THE INVERSE z-TRANSFORM

The z-transform is *unique*, or one-to-one. That is, a given sequence $x(n)$ will have a unique transform $X(z) = \mathcal{Z}[x(n)]$ and $X(z)$ will have a unique inverse $\mathcal{Z}^{-1}[X(z)]$ equal to $x(n)$. While the z-transform is generally straightforward to evaluate using the defining equation (3.5), it can be much more difficult to find a sequence given its ZT. In this section, we look at three methods for finding the inverse z-transform (IZT).

3.9.1 Power series method

$X(z)$ can sometimes be converted from closed form to infinite power series form using the Taylor series. However, the Taylor expansion is a power series, requiring positive powers of n, whereas $X(z)$ is an infinite series in z^{-1}. Rather than using a Laurent series expansion to cater for the negative powers, we simply make the substitution $g = z^{-1}$. Now, expanding $X(g)$ about point a using equations (3.6) and (3.7):

$$X(g) = \sum_{n=0}^{\infty} \frac{X^{(n)}(a)(g-a)^n}{n!} \tag{3.18}$$

The inverse can now be found by comparison with the defining summation (3.5).

A particularly useful special case of the Taylor series is the Maclaurin series for $a = 0$. To illustrate this, let us find the inverse of the transform $X(z) = \mathcal{Z}[u_n]$ which we obtained in example 3 of section 3.4 above. We found:

$$X(z) = \frac{z}{z-1} = \left(1 - z^{-1}\right)^{-1}$$

The Maclaurin series is found from equation (3.18) with $a = 0$:

$$X(g) = \sum_{n=0}^{\infty} \frac{X^{(n)}(0)g^n}{n!}$$

where $X(g) = (1-g)^{-1}$. It is easily shown that $X^{(n)}(0)$ equals $n!$, so that:

$$X(g) = \sum_{n=0}^{\infty} g^n$$

and

$$X(z) = \sum_{n=0}^{\infty} z^{-n}$$

This, of course, is exactly the geometric series with (complex) common ratio, g. By comparison with the defining equation (3.5), we see that $x(n)$ is simply equal to 1 for all $n \geq 0$, i.e. $x(n) = u_n$ as already known.

While evaluating the derivatives of $X(g)$ with respect to g at point $a = 0$ presents no problems, this is not true of the derivatives of $X(z)$ with respect to z. The latter are actually infinite, hence the necessity for the substitution.

3.9.2 Partial fraction expansion

In this technique, the rational polynomial $X(z)$ is expressed as a sum of simple terms, each with a readily identifiable inverse. As an example, suppose we wish to find the impulse response of a system, $h(n)$, given:

$$H(z) = \frac{120z^2 + 16z}{15z^2 + 2z - 1}$$

The roots of the denominator (the system poles) are $-1/3$ and $1/5$. Thus:

$$\begin{aligned} H(z) &= \frac{\left(8z + \frac{16}{15}\right) z}{\left(z + \frac{1}{3}\right)\left(z - \frac{1}{5}\right)} \\ &= \frac{Az}{\left(z + \frac{1}{3}\right)} + \frac{Bz}{\left(z - \frac{1}{5}\right)} \end{aligned}$$

It is now easy to show (e.g. using the 'cover-up rule') that $A = 3$ and $B = 5$, and hence:

$$H(z) = \frac{3z}{\left(z + \frac{1}{3}\right)} + \frac{5z}{\left(z - \frac{1}{5}\right)}$$

Comparing with example 4 of section 3.4 (with u_n implied for a causal system), the IZT is found to give:

$$h(n) = 3\left(-\frac{1}{3}\right)^n + 5\left(\frac{1}{5}\right)^n \qquad n \geq 0$$

3.9.3 Contour integral (residue) method

In section 3.2 above, we saw that the ZT of a sequence defines a Laurent series and, furthermore, the b_n coefficients of $(z - z_0)^{-n}$ can be found by contour integration. It happens that the required integration can be performed very simply using the 'theory of residues', and this gives us a powerful means of finding inverse z-transforms.

By definition of the ZT:

$$X(z) = \sum_{n=0}^{\infty} x(n) z^{-n}$$

which is a Laurent expansion of $X(z)$ about $z_0 = 0$. The corresponding Laurent series coefficients are (equation (3.9)):

$$\begin{aligned} b_n \ (= x(n)) &= \frac{1}{2\pi j} \oint_C X(z) z^{n-1} dz \qquad (n \geq 1) \qquad (3.19) \\ x_0 &= a_0 \end{aligned}$$

where C encloses the origin and its path lies entirely within the region of convergence R of $X(z)$. Because R is, in general, annular and centered on the origin (figure 3.2), there is no difficulty in finding such a C.

Since b_n here is identical to $x(n)$, equation (3.19) defines the IZT and forms a transform pair along with equation (3.5).

For the particular case of $n = 1$:

$$b_1 = \frac{1}{2\pi j} \oint_C X(z) dz$$

and:

$$\oint_C X(z) dz = 2\pi j b_1$$

which gives us a means of evaluating contour integrals from the Laurent expansion.

Now, the Laurent expansion is valid in each of two cases. First, if $X(z)$ is analytic at all points within C, then the expansion reduces to a Taylor series, all the b_n coefficients are zero, and the contour integral is also zero (in accordance with the Cauchy-Goursat theorem). In the other case, z_0 is an isolated singular point enclosed by C and b_1 is then non-zero. The term b_1 is called the *residue* of $X(z)$ at $z = z_0$, denoted $R_{z_0}[X(z)]$. When C encloses more than one singularity, each of them contributes to the contour integral by an amount equal to the corresponding residue. The *residue theorem* (whose proof is straightforward but not given here) states that for a contour C within which $F(z)$ is analytic except for a finite number (S) of singularities:

$$\oint_C F(z) dz = 2\pi j \sum_{k=0}^{S-1} R_{z_k}[F(z)]$$

Thus, the inverse z-transform in the form of equation (3.19) can be found from the residues of $X(z)z^{n-1}$ evaluated at the singularities of this function (which are, of course, identical to the singularities of $X(z)$ for $n \geq 1$):

$$x(n) = \sum_{k=0}^{S-1} R_{z_k}\left[X(z)z^{n-1}\right] \tag{3.20}$$

These residues can usually be evaluated without difficulty. Consider the case of rational $F(z)$ with a singularity (of order $p = 1$) at z_0. We can generally find a function ψ such that:

$$\psi(z) = F(z)(z - z_0)$$

where, unlike $F(z)$, $\psi(z)$ is analytic at z_0 because the $(z - z_0)$ term which makes $F(z)$ infinite at z_0 has been factored out. Provided $\psi(z_0) \neq 0$, it will have a Taylor expansion valid in $|z - z_0| < R$:

$$\psi(z) = \psi(z_0) + \frac{\psi'(z_0)}{1!}(z - z_0) + \frac{\psi''(z_0)}{2!}(z - z_0)^2 + \frac{\psi'''(z_0)}{3!}(z - z_0)^3 + \cdots$$

Hence, for $0 < |z - z_0| < R$:

$$F(z) = \frac{\psi(z_0)}{(z - z_0)} + \frac{\psi'(z_0)}{1!} + \frac{\psi''(z_0)}{2!}(z - z_0) + \frac{\psi'''(z_0)}{3!}(z - z_0)^2 + \cdots$$

Comparing with the Laurent expansion about z_0 using equation (3.8):

$$\begin{aligned} F(z) &= a_0 + a_1(z - z_0) + a_2(z - z_0)^2 + \cdots \\ &\quad + b_1(z - z_0)^{-1} + b_2(z - z_0)^{-2} + \cdots \end{aligned}$$

we see that the residue b_1 is equal to $\psi(z)$ evaluated at the singularity z_0. Thus, when $p = 1$, the terms in equation (3.20) are found as:

$$R_{z_k}\left[X(z)z^{n-1}\right] = (z - z_k)X(z)z^{n-1}\Big|_{z=z_k} \quad (3.21)$$

By a similar argument, it is easily shown that when $F(z)$ has a singularity of order p (which may be greater than 1) at z_0:

$$b_1 = \frac{\psi^{(p-1)}(z_0)}{(p - 1)!}$$

Thus, the residue of $X(z)z^{n-1}$ at a pole $z = z_k$ of order p is given more generally by:

$$R_{z_k}\left[X(z)z^{n-1}\right] = \frac{d^{p-1}}{dz^{p-1}}\left(\frac{(z - z_k)^p}{(p - 1)!}X(z)z^{n-1}\right)\Bigg|_{z=z_k} \quad (3.22)$$

As an example of the use of residues, let us find the impulse response $h(n)$ of the system studied in the previous chapter having difference equation (2.2):

$$y_n = 0.1x_n + 0.4x_{n-1} + 0.5y_{n-1}$$

Taking z-transforms:

$$Y(z) = 0.1X(z) + 0.4z^{-1}X(z) + 0.5z^{-1}Y(z)$$

Hence:

$$H(z) = \frac{Y(z)}{X(z)} = 0.1\left(\frac{z+4}{z-0.5}\right)$$

$H(z)$ has a single, real pole at $z = 0.5$ and the region of convergence is $|z| > 0.5$. From equation (3.20) with $S = 1$ singularities, $h(n)$ is equal to the residue of $H(z)z^{n-1}$ at $z = 0.5$. Since the order of the pole is $p = 1$, we can use the simpler form of equation (3.21) rather than the more general form of (3.22) to give:

$$\begin{aligned} h(n) &= (z-0.5)\left(0.1\frac{z+4}{z-0.5}\right)z^{n-1}\Big|_{z=0.5} \\ &= 0.1\left(z^n + 4z^{n-1}\right)\Big|_{z=0.5} \\ &= 0.1(0.5)^n + 0.4(0.5)^{n-1} \qquad n \geq 1 \end{aligned}$$

exactly as found in the previous chapter, but with use of the ZT and residues here greatly easing the task.

Note that this solution is only valid for $n \geq 1$, as required in equation (3.19). It is left as an exercise for the reader to show that $h(0)$ $(= a_0) = 0.1$.

As a further example, let us find the impulse response corresponding to the system function considered above in 3.9.2:

$$H(z) = \frac{\left(8z + \frac{16}{15}\right)z}{\left(z+\frac{1}{3}\right)\left(z-\frac{1}{5}\right)}$$

$H(z)$ has two poles, z_1 at $z = -1/3$ and z_2 at $z = 1/5$, each of order $p = 1$. From equation (3.20) with $S = 2$, and using the abbreviated notation R_{z_k} for $R_{z_k}[H(z)z^{n-1}]$:

$$h(n) = R_{-\frac{1}{3}} + R_{\frac{1}{5}}$$

Since $p = 1$ (for both residues), we use equation (3.21) to give:

$$\begin{aligned} R_{-\frac{1}{3}} &= \left(z+\frac{1}{3}\right)\frac{\left(8z+\frac{16}{15}\right)z}{\left(z+\frac{1}{3}\right)\left(z-\frac{1}{5}\right)}z^{n-1}\Bigg|_{z=-\frac{1}{3}} \\ &= z^n\left(\frac{8z+\frac{16}{15}}{z-\frac{1}{5}}\right)\Bigg|_{z=-\frac{1}{3}} \end{aligned}$$

$$= \left(-\frac{1}{3}\right)^n \left(\frac{-\frac{8}{3} + \frac{16}{15}}{-\frac{1}{3} + \frac{1}{15}}\right)$$

$$= 3\left(-\frac{1}{3}\right)^n$$

Similarly:

$$R_{\frac{1}{5}} = 5\left(\frac{1}{5}\right)^n$$

Hence:

$$h(n) = 3\left(-\frac{1}{3}\right)^n + 5\left(\frac{1}{5}\right)^n$$

as before.

Strictly, as stated for the previous example, this solution is only valid for $n \geq 1$, although it happens to give the correct $h(0)$ in this case.

Comparing this method with the previous one, we see that each of the partial fractions can be directly identified with one the residues. This observation gives some insight into how and why the cover-up rule works.

3.10 TIME RESPONSE OF A DT SYSTEM

The output from a discrete-time system with input $x(n)$ and impulse response $h(n)$ is:

$$y(n) = \mathcal{Z}^{-1}[Y(z)] = \mathcal{Z}^{-1}[H(z).X(z)]$$

So, the IZT gives us a way to find the time response of an LTI system for any input $x(n)$.

As we have seen, the IZT of a general function $F(z)$ is determined by the residues of $F(z)z^{n-1}$ at the poles of the function. Hence, the time response of a system depends primarily upon the poles of $H(z)$ and $X(z)$.

Consider the part of the output $y(n)$ due to the poles of $H(z)$ alone. Recall that $H(z)$, or equivalently $h(n)$, embodies all that we need to know about an LTI system. In the previous chapter, we defined the natural – or transient – response as that part of the output that was characteristic of the system and independent of any particular input. Hence, the part of the response due to the poles of $H(z)$ is in fact the natural response. Likewise, the part of the response due to the poles of $X(z)$ is identical to the forced response of the system.

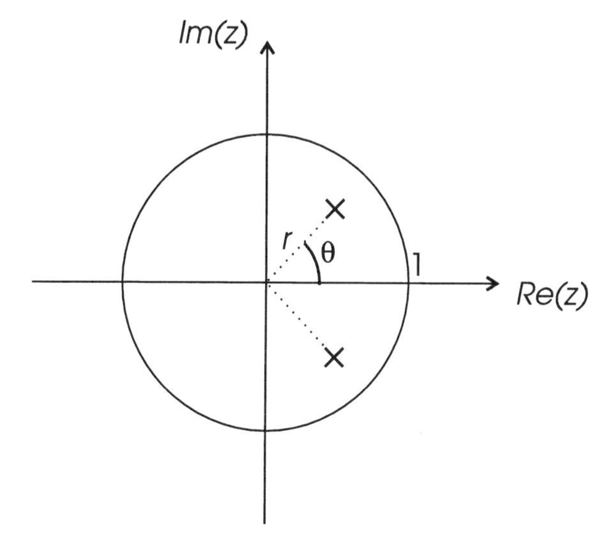

Fig. 3.13 Pole-zero diagram for a second-order DT system having a single pair of complex-conjugate poles.

It is perhaps worth making a further point about the impulse response, $h(n)$. In this case, the input is of course $x(n) = \delta_n$. Now, since $\mathcal{Z}[x(n) = \delta_n] = 1$, it is quite immaterial whether we assign this to $H(z)$ or to $X(z)$. This fact gives some insight into the problem we had in the previous chapter (section 2.5.1) when solving a difference equation for the input $x(n) = \delta_n$. (Recall that one difficulty was deciding whether to include the delta functions in the complementary solution or the particular solution.)

Let us consider the natural response in some detail. Take the simple yet general case of a second-order system with a single pair of complex-conjugate poles at $z = re^{\pm j\theta}$, as shown in figure 3.13. Thus:

$$H(z) = \frac{1}{\left(z - re^{j\theta}\right)\left(z - re^{-j\theta}\right)} \tag{3.23}$$

Using the residue method:

$$h(n) = R_{z=re^{j\theta}} + R_{z=re^{-j\theta}}$$

The first residue is given by:

$$
\begin{aligned}
R_{re^{j\theta}} &= \left(z - re^{j\theta}\right) \left. \frac{z^{n-1}}{\left(z - re^{j\theta}\right)\left(z - re^{-j\theta}\right)} \right|_{z=re^{j\theta}} \\
&= \frac{\left(re^{j\theta}\right)^{n-1}}{\left(re^{j\theta} - re^{-j\theta}\right)} \\
&= \frac{r^{n-1} e^{j(n-1)\theta}}{r\left(e^{j\theta} - e^{-j\theta}\right)} \\
&= \frac{r^{n-2} e^{j(n-1)\theta}}{\left(e^{j\theta} - e^{-j\theta}\right)} \\
&= \frac{r^{n-2} e^{j(n-1)\theta}}{2j \sin\theta}
\end{aligned}
$$

Similarly:

$$
R_{re^{-j\theta}} = -\frac{r^{n-2} e^{-j(n-1)\theta}}{2j \sin\theta}
$$

so that:

$$
\begin{aligned}
h(n) &= \frac{r^{n-2}}{2j \sin\theta} \left(e^{j(n-1)\theta} - e^{-j(n-1)\theta}\right) \\
&= \left(\frac{r^{n-2}}{\sin\theta}\right) \sin(n-1)\theta
\end{aligned}
\tag{3.24}
$$

Thus, the impulse response is oscillatory: the second-order system with complex-conjugate poles has resonant properties. In interpreting equation (3.24), note that the $\sin\theta$ denominator is (because θ is a constant) a simple constant. The $\sin(n-1)\theta$ term, however, has a dependence upon n, i.e. upon (discrete) time, so that the response is sinusoidal with frequency dependent upon θ. Further, its magnitude has a time dependence on the factor r – the distance of the poles from the z-plane origin. For $r < 1$, $h(n)$ reduces exponentially with time, n. For $r \geq 1$, however, it grows in unbounded fashion as $n \to \infty$ and the system is hence unstable. This leads us to the very important criterion – mentioned several times previously – that in order to be stable, a DT system must have all its poles within the z-plane unit circle. As a consequence of $h(n)$ depending only on the residues of $H(z)z^{n-1}$ at the system poles, there is no such restriction on the system zeros for stability to hold.

Note that the impulse response $h(n)$ as expressed in equation (3.24) does not start at $n = 0$, but at $n = 2$ (see exercise 3.4). This is because the system has more poles than zeros – the two poles of figure 3.13 are not balanced by a corresponding number of zeros at the origin, as required for $H(z)$ in equations (3.14) and (3.15) to be a rational polynomial in z^{-1}. The impulse response can be made to start at $n = 0$ by multiplying $H(z)$ by z^2 (adding a second-order zero at the origin) so as to balance the number of poles and zeros. This is an instance of the right-shift property: multiplying by z^k will shift the sequence right by $-k$ samples, i.e. k samples *leftwards*, when k is positive.

It is left as an exercise for the reader to show that multiplying $H(z)$ by z^2, i.e. adding two zeros at the origin, will have the effect of replacing n in equation (3.24) by $(n+2)$ to give:

$$h(n) = \left(\frac{r^n}{\sin\theta}\right)\sin(n+1)\theta$$

Figure 3.14(a) shows the impulse response plotted for the two representative cases of $r = 0.9$, $\theta = \pi/6$ rad and $r = 0.8$, $\theta = \pi/12$ rad. As can be seen, the impulse response is much more oscillatory in the former case, and several periods of oscillation occur before the transient dies away to yield the steady-state response of zero. When θ is halved (so that the poles move closer to the positive real z-axis) and r reduced from 0.9 to 0.8, the oscillatory transient is seen to double in period and is very considerably reduced in amplitude.

Consider the step response, $s(n)$, of this system. Exercise 3.16 shows that this is given as:

$$s(n) = \frac{r^{n+2}\sin(n+1)\theta - r^{n+1}\sin(n+2)\theta + \sin\theta}{\sin\theta\left(1 - 2r\cos\theta + r^2\right)}$$

as depicted in figure 3.14(b) for the same r and θ values as previously. Again, the response is much more oscillatory for the case of $r = 0.9$ and $\theta = \pi/6$ rad. As before, halving θ and reducing r doubles the period of the oscillatory transient and reduces it markedly in amplitude.

This behaviour can be understood from the pole-zero diagram of figure 3.13. Application of the step-function input amounts to input of a wideband signal (in which all frequencies are present), which excites the system into resonance. The frequency of oscillation is determined by the angle θ $(= \omega_{osc}T)$ since it is at this frequency that a general point on the unit circle, $e^{j\omega T}$, is closest to the pole with positive θ. For instance, from figure 3.14(a) and (b), we see that when $\theta = \pi/6$ rad, the period of oscillation, P, is 12 samples. This arises from:

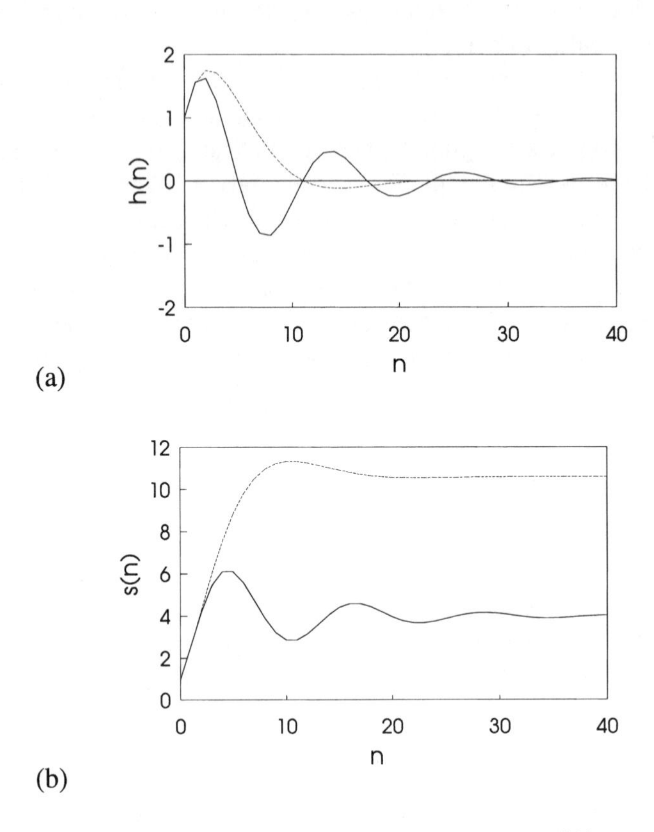

Fig. 3.14 Time responses of second-order DT system with poles at $re^{\pm j\theta}$: (a) impulse responses for $r = 0.9$, $\theta = \pi/6$ rad (full line) and for $r = 0.8$, $\theta = \pi/12$ rad (dotted line) (b) step responses for same values of r and θ.

$$\begin{aligned} \omega_{osc}T &= 2\pi f_{osc}T \\ &= 2\pi T/P \\ \text{so that } 2\pi T/P &= \pi/6 \\ \text{and} \quad P &= 12T \end{aligned}$$

with T (because n values naturally increment by 1) implicitly equal to 1 sample.

The closer the poles are to the unit circle, i.e. the nearer r is to 1, the larger will be the amplitude of the oscillatory response. When $r = 1$ (poles on the unit circle), the response is purely oscillatory, and at the natural frequency, ω_n, of the resonant system.

For $r < 1$, damping is introduced and $\omega_{osc} < \omega_n$. For both conditions shown in the figure, the system is under-damped because the poles are a complex-conjugate pair. In fact, the critically-damped condition corresponds to coincident poles on the real axis, while over-damping occurs when the poles are real but different.

It is tempting to assume from this explanation that r is the sole determinant of the system's damping factor, ζ, and that, likewise, θ is the sole determinant of the natural frequency of the system, ω_n. In fact, the relation between them is more complex than this. From exercise 3.17, we find that for complex-conjugate poles:

$$\begin{aligned} r &= e^{-\zeta\omega_n T} \\ \theta &= \omega_n T\sqrt{1-\zeta^2} \end{aligned}$$

Hence, r and θ between them jointly determine ζ and ω_n. When $r = 1$, then $\zeta = 0$ and $\omega_{osc} = \omega_n$, as described above. At the other extreme, when $\theta = 0$ (so that the complex-conjugate poles become coincident real poles), $\zeta = 1$ and the time response is non-oscillatory ($\omega_{osc} = 0$).

3.11 SUMMARY

We have covered a good deal of very important ground in this chapter. A sound understanding of the properties of discrete-time signals and systems depends on a good grasp of the concepts underlying the z-transform. Starting from the definition of the ZT as an infinite series in z^{-1}, where $z = e^{sT} = e^{(\sigma+j\omega)T}$, we interpreted z^{-1} as a unit-delay or shift operator applied to a sequence.

The ZT of a sequence $x(n)$, $X(z)$, was then shown to be a Laurent series expansion about the z-plane origin, $z_0 = 0$. The Laurent series is a generalisation of the Taylor series which (unlike the Taylor series) can be used at a point z_0 when that point is a singularity. Since Laurent series are well-studied, many standard results from complex analysis are applicable to the ZT. In particular, the infinite series expression for particular ZTs can be summed to give a closed-form expression, and the region of convergence found.

We then explored the relation of the ZT to the Fourier transform of a sequence. The particular FT representation used was obtained from the Fourier integral by numerical integral using the rectangular approximation studied in chapter 2. This representation is continuous (rather than discrete) in frequency. It was found that the FT (like the Laplace transform) of a signal sampled with sampling period T repeats indefinitely in frequency with period $2\pi/T$. By contrast, because the simple complex variable z in the ZT replaces the complex exponential $e^{j\omega T}$ in the FT, the ZT is non-periodic in

z. Apart from its non-periodicity, the important properties of linearity, right-shift and convolution were detailed and proved. Also, because the ZT of a sequence involves a convergence factor ($e^{\sigma T}$) which is absent from the FT, the former can converge when the latter does not. An example of this was seen in the case of the unit-step sequence, u_n.

The link between signals and systems was made by considering the ZT of the 'signal' $h(n)$. When this is the impulse response of a DT system, $\mathcal{Z}[h(n)] = H(z)$ is the system function. Starting from the general difference equation introduced in the previous chapter, $H(z)$ was shown to be a rational polynomial in z^{-1} with coefficients identical to those appearing in the difference equation. The function $H(z)$ is of primary importance in the study of DT systems, being a complete description of the properties of a given system (as, of course, is its inverse transform $h(n)$).

One very useful way to visualise the properties of a system is to plot the poles and zeros of $H(z)$ on the (complex) z-plane. The system poles are those values of z which make $H(z)$ go infinite, while the zeros are those values of z which make $H(z)$ zero. Because $H(z)$ is a rational polynomial in z^{-1}, poles and zeros must either be real or occur in complex-conjugate pairs. By considering the mapping between the variable s of the Laplace transform and z, we showed that constant frequency in z corresponds to lines of constant slope terminating at the z-plane origin, while constant σ corresponds to circles centred on the z-plane origin. In particular, $\sigma = 0$ corresponds to the z-plane unit circle. Since $\sigma = 0$ for the FT, the FT of a sequence can be found by evaluating its ZT around the z-plane unit circle defined by the locus of points $e^{j\omega T}$.

We next considered the response of a system to a complex exponential input, $Ae^{j\omega T}$, which is an eigenfunction of a general linear system. Thus, the magnitude of this particular input is modified by the factor $|H(e^{j\omega T})|$ and its phase shifted by $\angle H(e^{j\omega T})$. Hence, $H(e^{j\omega T})$ is the eigenvalue of the system, more specifically called the frequency response. It is formally identical to the FT of the impulse response sequence, h_n, i.e. to the evaluation of $H(z)$ around the z-plane unit circle. Consequently, $H(z)$ must converge around the unit circle for the corresponding system to be stable. The magnitude of this complex variable is the magnitude response, $M(\omega T) = |H(e^{j\omega T})|$, and its phase is the phase response, $\Phi(\omega T) = \angle H(e^{j\omega T})$. The magnitude and phase responses for a particular system can, therefore, be found algebraically by substituting $e^{j\omega T}$ for z in the system function and evaluating the modulus and argument respectively. An alternative is the graphical approach which determines the frequency response from the modulus and phase of vectors drawn from the z-plane poles and zeros to the point $e^{j\omega T}$ on the unit circle. The graphical approach makes it rather easier to visualise the magnitude response of a particular system, but this is less true of the phase response. However, the approach did give insight into the phase ambiguities which arise when $\Phi(\omega T)$ is evaluated using the inverse tangent function. We saw too why zeros and poles at the

z-plane origin have no effect on the magnitude response while they do affect the phase response. Further, the graphical method aids understanding of the periodicity and symmetry properties of $M(\omega T)$ and $\Phi(\omega T)$. Both responses are periodic in ω with period $2\pi/T$: the former is even symmetric about the folding (Nyquist) frequency π/T while the latter is odd symmetric about this frequency. The symmetry properties were explained on the basis that the system's z-plane poles and zeros are either real or occur in complex-conjugate pairs. Later on (chapter 6) when we study the Fourier representation of DT signals in some detail, we will gain additional and more general insight into these properties.

Having considered the frequency response of a DT system, it was necessary to postpone discussion of the time response until we had dealt with the inverse z-transform (IZT). We introduced three methods for finding the IZT: the power series method, partial fraction expansion and the contour integral (or residue) method. Of these, the latter is the most general and convenient in practice, although its use does demand a proper understanding of some topics in complex analysis.

We then discussed the time response of a DT system as the inverse z-transform of $Y(z) = H(z).X(z)$. Because the IZT of a general function $F(z)$ depends primarily on the residues of $F(z)z^{n-1}$, the system zeros have only a secondary effect on the time response. The part of $y(n) = \mathcal{Z}^{-1}Y(z)$ due to the poles of $H(z)$ is the natural response of the system, while the part due to the poles of $X(z)$ (i.e. due to the particular input sequence) is the forced response. By considering the general case of a complex-conjugate pair of poles, we showed that the time response to the unit-sample input (i.e. the impulse response) is oscillatory. That is, the second-order system with complex-conjugate poles has resonant properties. However, the oscillation vanishes when the poles are real. Furthermore, the amplitude of oscillation grows with time if the poles are outside the z-plane unit circle, but decays with time if they are within it. Because all poles of a system with a real impulse response are either complex-conjugate pairs or real, this leads to the outstandingly important stability criterion of a DT system: namely, all poles must lie within the unit circle. There is no such condition relating to the system zeros.

We showed that the effect of an imbalance in the number of poles and the number of zeros is to shift the time response, i.e. an equal number of poles and zeros is necessary for $h(n)$ to start at $n = 0$. This was seen to be a consequence of the right-shift property. Finally, we interpreted the oscillatory response of the second-order system with complex-conjugate poles in terms of its pole-zero diagram, showing how the familiar concepts of damping and natural frequency for a CT second-order system related to the DT case.

3.12 EXERCISES

3.1 Evaluate the z-transforms of the following and state the region of convergence:

(i) the rectangular pulse, $u(n) - u(n-p)$

(ii) Ae^{-anT}

(iii) $e^{-nT} + 2e^{-2nT}$

3.2 Show that the z-transform of the sequence $nx(n)$ is given by $-z\frac{dX(z)}{dz}$ and give the region of convergence.

3.3 We have seen that convolution in the discrete-time domain corresponds to multiplication in the complex-frequency (z) domain. In this exercise, we consider the dual of this convolution property.

Suppose sequences $x(n)$ and $y(n)$ are multiplied to give:

$$w(n) = x(n).y(n)$$

Show that:

$$W(z) = \frac{1}{2\pi j}\oint_C Y\left(\frac{z}{\nu}\right) X(\nu)\nu^{-1}d\nu$$

and state the region of convergence.

By the changes of variable:

$$\nu = e^{j\phi} \text{ and } z = e^{j\theta}$$

show that $W(e^{j\theta})$ can be expressed as a convolution integral carried out around the circle $e^{j\theta}$ in the z-plane. (For this reason, the process is referred to as circular convolution.)

3.4 Given the (causal) sequence $x(n) = \{x_0, x_1, \ldots\}$, prove the *initial value theorem*:

$$x_0 = \lim_{z \to \infty} X(z)$$

Consider the system whose system function is given by equation (3.23). Use the theorem to show that the h_0 and h_1 terms of the impulse response are zero. Also verify that the h_2 and h_3 terms found using the theorem are consistent with the values obtained using equation (3.24) directly.

3.5 What is the implication of the initial value theorem for the convergence of a causal sequence at infinity? Taking this result, together with the fact that $H(z)$ must converge on the unit circle for the corresponding system to be stable, show that a stable system must have all its poles within the z-plane unit circle.

3.6 Given the *final value theorem*:

$$\lim_{n\to\infty} x(n) = \lim_{z\to 1}\left(\frac{z-1}{z}\right) X(z)$$

show that the steady-state response of a DT system to the unit-step input, $y_{ss}(n)$, can be found by setting z to 1 in $H(z)$. Hence, show that $y_{ss}(n)$ can also be found from the zero-frequency response.

3.7 In speech processing, it is common to *pre-emphasise* the speech signal before further processing. This boosts the weaker high frequencies at the expense of the stronger low frequencies. Pre-emphasis is achieved using the difference equation:

$$y_n = x_n - \alpha x_{n-1}$$

From the z-plane pole-zero-diagram, sketch out the frequency responses (magnitude and phase). What would appear to be a suitable range of values for α?

3.8 Evaluate the frequency response (magnitude and phase) of the IIR system considered in the previous chapter, whose difference equation (2.2) is:

$$y_n = 0.1x_n + 0.4x_{n-1} + 0.5y_{n-1}$$

both algebraically and graphically.

What is the step response of this system?

3.9 Evaluate the frequency response (magnitude and phase) of the IIR system whose difference equation is:

$$y_n = 0.5x_n + 0.2y_{n-1}$$

What is the response of this system to the input $x(n) = \sin(n\omega T)$?

3.10 Find the frequency responses of the following differentiators studied in the previous chapter:

(i) $y_n = \frac{1}{T}(x_n - x_{n-1})$ (causal version of equation (2.7))

(ii) $y_n = \frac{1}{2T}(3x_n - 4x_{n-1} + x_{n-2})$ (equation (2.14))

What would the ideal differentiator response be?

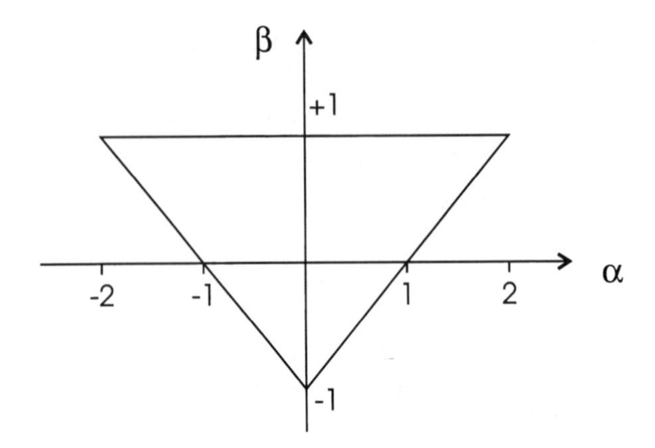

Fig. 3.15 Stability triangle for second-order system with quadratic coefficients α and β.

3.11 Find the frequency responses of the following integrators studied in the previous chapter:

(i) $y_n = y_{n-1} + x_n T$ (rectangular integrator)

(ii) $y_n = y_{n-2} + \frac{T}{2}(x_{n-1} + x_n)$ (trapeziodal integrator)

(iii) $y_n = y_{n-2} + \frac{T}{3}(x_{n-2} + 4x_{n-1} + x_n)$ (piecewise-quadratic integrator)

What would the ideal integrator response be?

3.12 Show that, with one exception, the zeros of an M-point moving-average filter are the M roots of unity. Which zero is the exception? Why is it an exception?

3.13 The function arctan2( ) takes two arguments, namely the (signed) lengths of the opposite and the adjacent sides of a right-angled triangle, and returns the value of the appropriate angle. Does the use of this function avoid the problems of phase ambiguity inherent in the use of the inverse tangent, $\tan^{-1}$, in finding the phase response of a system?

3.14 Consider the second-order system defined by:

$$H(z) = \frac{1}{(1 + \alpha z^{-1} + \beta z^{-2})}$$

Show that, in order for the system to be stable, the quadratic coefficients α and β must lie within the triangular region shown in figure 3.15. Also find the boundary within this *stability triangle* between the regions in which real and complex poles are located.

3.15 (a) Equation (3.24) describes the impulse response for the second-order system having two complex-conjugate poles, at $z = re^{\pm j\theta}$, studied in section 3.10. Use the small-angle formula for sine to evaluate $h(n)$ when the poles are coincident on the real axis ($\theta = 0$). Is the response still oscillatory? Must the poles still be within the unit circle for stability?

(a) By writing a computer program, using a spreadsheet, or otherwise, graph the magnitude response $|H(e^{j\omega T})|$ for the cases:

(i) $r = 0.9, \theta = \pi/6$ rad;

(ii) $r = 0.8, \theta = \pi/12$ rad.

Relate the frequency of the resonant peak seen in the magnitude response to the value of θ.

3.16 Show that, for general values of r and θ, the step response of the second-order system of the previous example is:

$$s(n) = \frac{r^{n+2}\sin(n+1)\theta - r^{n+1}\sin(n+2)\theta + \sin\theta}{\sin\theta\left(1 - 2r\cos\theta + r^2\right)}$$

What is the frequency of oscillation?

Separate $s(n)$ out into a particular solution (having the same time dependence as the input) and a complementary solution (depending on the inherent properties of the system). By considering the limit of $s(n)$ in the above expression as $n \to \infty$, obtain a value for the steady-state response in terms of r and θ and show that it correctly predicts the values seen in figure 3.14(b).

Compare these with the values found from:

(i) the final value theorem;

(ii) the zero-frequency response of the system.

3.17 (a) The general differential equation for a second-order continuous-time system is:

$$\frac{d^2y}{dt^2} + 2\zeta\omega_n\frac{dy}{dt} + {\omega_n}^2 y = kx$$

where ζ and ω_n are the damping factor and natural frequency respectively.

Determine the s-plane pole positions for the under-damped condition, $\zeta^2 < 1$.

(b) Show that the corresponding under-damped discrete-time system has poles at $z = re^{\pm j\theta}$, where:

$$r = e^{-\zeta\omega_n T}$$

$$\theta = \omega_n T\sqrt{1-\zeta^2}$$

Hence, find the damping factor and natural frequency for the case of $r = 0.9$ and $\theta = \pi/6$ rad. Also find expressions for ω_n and ζ in terms of r and θ.

3.18 What is meant by a *conformal* transformation? Is the z-transform conformal? What is the relevance of this to discrete-time signals and systems?

4

Infinite impulse response (IIR) filters

Given any system function $H(z)$ describing a DT system, we can always find the magnitude and phase responses, $M(\omega T) = |H(e^{j\omega t})|$ and $\Phi(\omega T) = \angle H(e^{j\omega t})$ respectively, as described in the previous chapter. However, the converse is not true. There is unfortunately no general theory for taking arbitrary magnitude and phase responses and synthesising an appropriate $H(z)$ from them.

By contrast, 'analogue' synthesis procedures for CT signal-processing systems – or filters – do exist. There is, in principle, no difficulty in designing an analogue filter to a particular specification. One useful approach to the design of DT filters is, therefore, to model the desired system as a CT filter and then convert it by some means into DT form. That is, we have some analogue system described by the function $H(s)$ which acts as the *prototype* for a 'digital' filter to be implemented as a DT system described by the function $H(z)$.

In this chapter, two powerful techniques for transforming $H(s)$ to $H(z)$ are considered:

- the bilinear transform;
- the impulse invariance method.

Both methods produce a recursive (infinite impulse response) realisation.

4.1 THE BILINEAR TRANSFORM

The starting point for our design is a prototype analogue (CT) filter described by its system function, $H(s)$, which we wish to convert it to a digital (DT) filter having the same magnitude response.

So, can we use the now-familiar $s \to z$ mapping defined by $z = e^{sT}$? In this case, a stable prototype will transform to a stable digital filter. For instance, a real left-half s-plane pole at $s = -a$ will map to a real pole at $z = e^{-aT}$ within the unit circle. The problem, however, is that the DT realisation will have a periodic magnitude response. Recall that to find the magnitude response, we put $z = e^{j\omega T}$ in $H(z)$, i.e. we evaluate $H(z)$ around the unit circle. As ωT goes from 0 to ∞, we go round and round the unit circle in the z-plane with period $\omega T = 2\pi$. This will lead to a problem of aliasing, just like that arising when we undersample a continuous-time signal. The response of the analogue prototype extends up to infinite frequency, but application of the z-transform will force the response of the digital realisation to 'fold' around the Nyquist frequency, $\omega_N = \omega_s/2 = \pi/T$.

The solution to this problem is to find a new transform which retains all the mapping properties of the z-transform, namely:

- the imaginary axis of the s-plane maps to the unit circle;
- the left half-plane of the s-plane maps to the inside of the unit circle;

except that as frequency goes from $-\infty$ to ∞, we go *once only* around the z-plane unit circle, i.e. ωT goes from $-\pi$ to π. Essentially, we require a two-step transform. The first step maps the entire s-plane to the strip defined by $-\pi/T \leq |Im(w)| \leq \pi/T$ in some intermediate complex domain w, as depicted in figure 4.1. This is referred to as *band-limiting*. The second step maps the left-half of the strip to the interior of the unit circle, and the right half to the exterior. The 'standard' z-transform can be used for the second step.

The resulting two-stage process is called the *bilinear* z-transform or, more usually, simply the *bilinear transform*.

A useful band-limiting function on which to base the transform is tanh( ), where the mapping is:

$$\begin{aligned} x &\to \tanh x \\ &= \left(\frac{e^x - e^{-x}}{e^x + e^{-x}}\right) \end{aligned}$$

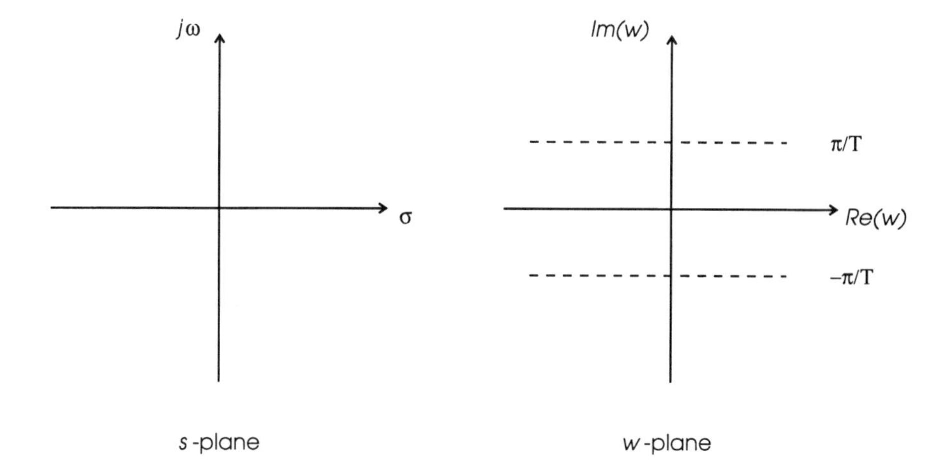

Fig. 4.1 The bilinear transform can be viewed as a two-step mapping in which the first step is a mapping of the entire s-plane to the strip $-\pi/T \leq |Im(w)| \leq \pi/T$ in the intermediate domain, w.

$$= \left(\frac{1-e^{-2x}}{1+e^{-2x}}\right)$$

Figure 4.2 depicts this mapping, and shows that it is band-limited to ± 1.

Now, as x goes from $-j\pi/2$ to $j\pi/2$ on the right-hand side, the left-hand side of the mapping function goes from $-\infty$ to ∞. This suggests we should replace x by $sT/2$:

$$\begin{aligned}\frac{sT}{2} &\rightarrow \left(\tanh\frac{sT}{2}\right)\\ &= \left(\frac{1-e^{-sT}}{1+e^{-sT}}\right)\end{aligned}$$

and the required s to w mapping is:

$$s \rightarrow w = \frac{2}{T}\left(\frac{1-e^{-wT}}{1+e^{-wT}}\right)$$

Combining this band-limited mapping with $z = e^{wT}$ will then map the $\pm j\pi/T$ strip to the inside of the unit circle. Hence, the bilinear transform is:

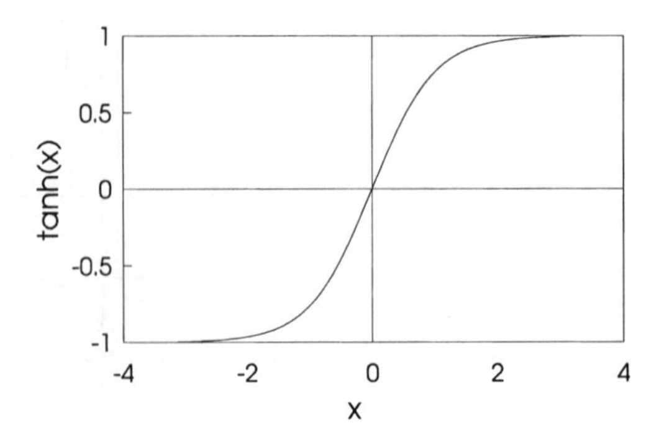

Fig. 4.2 The tanh function, which forms the basis of the bilinear transform, is band-limited to ± 1.

$$s \rightarrow \frac{2}{T}\left(\frac{1 - z^{-1}}{1 + z^{-1}}\right) \tag{4.1}$$

Because (unlike the 'standard' z-transform), this is a unique, one-to-one mapping, there is a unique inverse given by:

$$z \rightarrow \frac{\left(\frac{2}{T} + s\right)}{\left(\frac{2}{T} - s\right)} \tag{4.2}$$

4.1.1 Mapping properties

A direct consequence of the use of the band-limited tanh( ) function in the bilinear transform is that the frequency variables of the CT prototype and DT derivative are actually different. Loosely, these are referred to as 'analogue' and 'digital' frequency, respectively. We use Ω to denote analogue frequency and ω to denote digital frequency.

So, in equation (4.2):

$$z \rightarrow \frac{\frac{2}{T} + \sigma + j\Omega}{\frac{2}{T} - \sigma - j\Omega}$$

On the $j\Omega$ axis of s, $\sigma = 0$, and:

$$z \rightarrow \frac{\frac{2}{T} + j\Omega}{\frac{2}{T} - j\Omega}$$

$$|z| \;=\; 1$$

Thus, the $j\Omega$ axis maps to the unit circle in the (bilinear) z-plane, as desired. Also, as Ω varies from 0 to ∞, $\angle z$ goes from 0 to π radians; while as Ω varies from 0 to $-\infty$, $\angle z$ goes from 0 to $-\pi$ radians. A similar result holds for any constant value of σ. Hence, lines of constant σ in s map to circles in z which are traversed only once as Ω goes from $-\infty$ to ∞. In this way, the bilinear transform avoids the problem inherent in the standard z-transform, where the unit circle in z is traversed with period $2\pi/T$ as frequency varies, leading to aliasing of the frequency response.

Now suppose $\sigma < 0$ $(= -k$, say):

$$z \;\rightarrow\; \frac{\frac{2}{T} - k + j\Omega}{\frac{2}{T} + k - j\Omega}$$

$$|z| \;\rightarrow\; \sqrt{\frac{\frac{4}{T^2} - \frac{2k}{T} + k^2 + \Omega^2}{\frac{4}{T^2} + \frac{2k}{T} + k^2 + \Omega^2}} < 1$$

That is, left-half poles in s map to the interior of the unit circle in the (bilinear) z-plane. Hence, the same stability criteria apply as for the conventional $s \rightarrow z = e^{sT}$ mapping. Stable analogue prototype filters therefore yield stable DT derivatives.

4.1.2 Illustrative examples

Some very important points about the bilinear transform can be illustrated by considering two simple examples.

Example 1: An ideal CT differentiator has system function $H(s) = s$ (see page 33 of section 1.10). Given this prototype, obtain a DT realisation. Compare the magnitude responses of the CT and DT systems assuming the sampling period, T, is equal to 1 second.

Applying the bilinear transform, equation (4.1), directly:

$$H(z) = \frac{2}{T}\left(\frac{1 - z^{-1}}{1 + z^{-1}}\right) \tag{4.3}$$

Comparing the system function (equation (3.14)) with the general difference equation (2.5), the difference equation of the differentiator is found as:

$$y_n = \frac{2}{T}\,(1 - x_{n-1} - y_{n-1})$$

From equation (4.3), $H(z)$ has a zero at $z = 1$ which can be clearly identified via the inverse transform, equation (4.2), with the zero of $H(s)$ at the s-plane origin. The

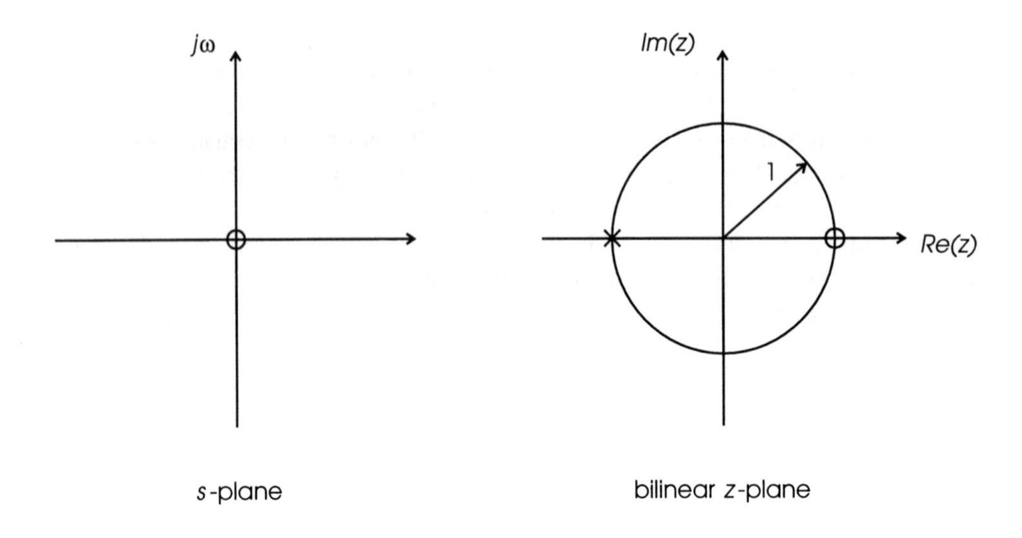

Fig. 4.3 The ideal differentiator $H(s) = s$ has a single zero at the s-plane origin. The bilinear transform maps this to a zero at $z = 1$ in the DT realisation, which also has a finite pole at $z = -1$ corresponding to the pole of $H(s)$ at infinity.

bilinear transform maps the s-plane origin to the $z = 1$ point, just as the standard z-transform does. However, $H(z)$ has a pole at $z = -1$ that $H(s)$ appears to lack. In fact, $H(s)$ does have a pole but it is at infinity, and the bilinear transform maps this to the point $z = -1$ (i.e. $\omega T = \pi$ rad) as we have seen above. In general, $H(s)$ and the $H(z)$ derived from it have the same numbers of poles and zeros, provided poles/zeros at $\pm\infty$ in s are taken into account. Figure 4.3 depicts the pole-zero diagrams for both $H(s)$ and $H(z)$.

So, although $H(s)$ had no finite poles, $H(z)$ does have. In general, any $H(s)$ prototype without finite poles will transform to a DT realisation having such poles. Hence, the bilinear transform always produces an IIR system, since the finite poles in $H(z)$ relate to feedback (recursive) terms in the difference equation. However, we know from chapter 2 that a differentiator should be an FIR system, so we have a clear expectation that the DT differentiator of equation (4.3) will be less that ideal.

Let us compare the magnitude responses of $H(s)$ and $H(z)$. For the CT prototype:

$$\begin{aligned} H(j\Omega) &= j\Omega \\ \text{and} \quad |H(j\Omega)| &= \Omega \end{aligned}$$

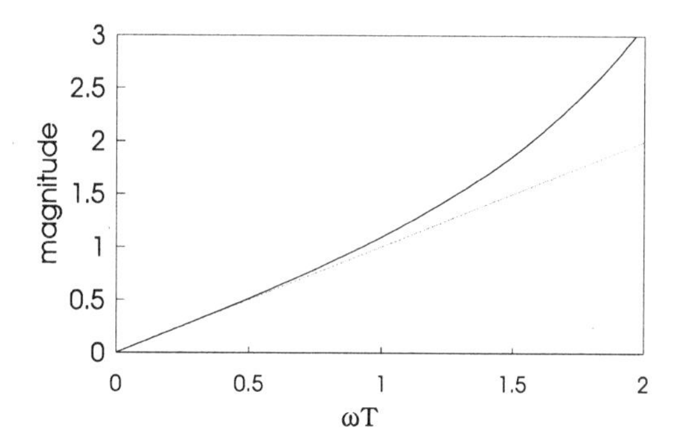

Fig. 4.4 Magnitude response of DT differentiator produced from $H(s) = s$ prototype using the bilinear transform (full line) and $T = 1$ second. The dotted line depicts the ideal magnitude response (proportional to frequency) for the analogue prototype differentiator.

Thus, the magnitude response is proportional to (analogue) frequency, being zero at zero frequency, as it should be for the ideal differentiator.

For the DT realisation:

$$\begin{aligned} H\left(e^{j\omega T}\right) &= \frac{2}{T}\left(\frac{1-e^{-j\omega T}}{1+e^{-j\omega T}}\right) \\ &= \frac{2}{T}\left(\frac{e^{\frac{j\omega T}{2}}-e^{\frac{-j\omega T}{2}}}{e^{\frac{j\omega T}{2}}+e^{\frac{-j\omega T}{2}}}\right) \\ &= \frac{2}{T}j\left(\frac{\sin\frac{\omega T}{2}}{\cos\frac{\omega T}{2}}\right) \\ \text{so that}\quad \left|H\left(e^{j\omega T}\right)\right| &= \frac{2}{T}\left|\tan\left(\frac{\omega T}{2}\right)\right| \\ &= \frac{2}{T}\tan\left(\frac{\omega T}{2}\right) \qquad (0 \le \omega T < \pi) \qquad (4.4) \end{aligned}$$

Figure 4.4 shows the magnitude response and its departure from the required proportionality to frequency as ωT increases from 0 up to 2 rad (and with T assumed equal to 1). As can be seen, the departure is not too serious at low frequencies, since for small ω:

$$\frac{2}{T}\tan\left(\frac{\omega T}{2}\right) \approx \frac{2}{T}\left(\frac{\omega T}{2}\right) = \omega$$

At $\omega T = 1$ rad, the deviation from unit magnitude is about 9%. For frequencies much above this, however, there is rapid divergence from proportionality, i.e. from the desired magnitude response. For example, when $\omega T/2 = \pi/4$, the response will be $2/T$ but it ought to be $\omega = \pi/2T$. Indeed, because $\tan(x)$ goes infinite at $x = \pi/2$, the divergence is infinite at the folding frequency ($\omega T/2 = \pi/2$). This can also be seen from the pole-zero diagram (figure 4.3) where the pole of $H(z)$ on the unit circle at $z = -1$ leads to an infinite response at infinity in analogue frequency but at the folding frequency in digital frequency.

Note finally that both the CT and DT differentiators have the same phase response, i.e. a constant $\pi/2$ phase lead.

Example 2: convert the Butterworth first-order low-pass filter $H(s) = 1/(s+a)$ with cut-off frequency $\Omega_c = a$ to DT form. What is the cut-off frequency, ω_c, of the digital filter? Evaluate ω_c when $\Omega_c = 10\pi$ rad.s^{-1} and the sampling rate is 40 Hz (i.e. $T = 25$ ms).

Applying the bilinear transform, equation (4.1):

$$\frac{1}{s+a} \to \frac{1}{\frac{2}{T}\left(\frac{1-z^{-1}}{1+z^{-1}}\right)+a}$$

$$H(z) = \frac{(1+z^{-1})}{\frac{2}{T}(1-z^{-1})+a(1+z^{-1})}$$

$$= \frac{(1+z^{-1})}{\left(\frac{2}{T}+a\right)-\left(\frac{2}{T}-a\right)z^{-1}}$$

Thus, $H(z)$ has a single pole at:

$$z = \left(\frac{\frac{2}{T}-a}{\frac{2}{T}+a}\right)$$

corresponding to the single pole of $H(s)$ at $s = -a$. In similar fashion to the previous example, $H(z)$ has a single zero at $z = -1$ which corresponds to the infinite zero in $H(s)$.

The frequency response is:

$$
\begin{aligned}
H\left(e^{j\omega T}\right) &= \frac{\left(1+e^{-j\omega T}\right)}{\left(\frac{2}{T}+a\right)-\left(\frac{2}{T}-a\right)e^{-j\omega T}} \\
&= \frac{\left(e^{\frac{j\omega T}{2}}+e^{-\frac{j\omega T}{2}}\right)}{\left(\frac{2}{T}+a\right)e^{\frac{j\omega T}{2}}-\left(\frac{2}{T}-a\right)e^{-\frac{j\omega T}{2}}} \\
&= \frac{\cos\left(\frac{\omega T}{2}\right)}{\frac{2j}{T}\sin\left(\frac{\omega T}{2}\right)+a\cos\left(\frac{\omega T}{2}\right)} \\
&= \frac{1}{a+\frac{2j}{T}\tan\left(\frac{\omega T}{2}\right)}
\end{aligned}
$$

The cut-off frequency $\Omega_c = a$ for the CT prototype is replaced by ω_c in $H(z)$ where:

$$
\begin{aligned}
\frac{2}{T}\tan\left(\frac{\omega_c T}{2}\right) &= a \\
\omega_c &= \frac{2}{T}\tan^{-1}\left(\frac{aT}{2}\right)
\end{aligned}
$$

Given that $\Omega_c = a = 10\pi$ rad.s^{-1} and $T = 25$ ms:

$$
\begin{aligned}
\omega_c &= \frac{2}{25\times 10^{-3}}\tan^{-1}\left(5\pi\times 25\times 10^{-3}\right) \\
&= 29.93 \text{ rad.s}^{-1}
\end{aligned}
$$

(c.f. 31.42 rad.s^{-1} for the analogue prototype)

The reader is warned that it is very easy to become confused between the tan and $\tan^{-1}$ functions which appear above, according to the direction of the mapping. The simplest way to handle any such confusion is to remember that the bilinear transform maps infinite analogue frequency to the folding digital frequency. Hence, digital values are always lower than their corresponding analogue values. For instance, the digital cut-off value obtained in this example ($\omega_c = 29.93$ rad.s^{-1} or $f_s = 4.76$ Hz) is, of course, lower than the corresponding analogue cut-off ($\Omega_c = 31.42$ rad.s^{-1} or $f_s = 5$ Hz).

4.1.3 Filter design by 'pre-warping'

As we have seen, analogue and digital frequency are different, and the relation between them is non-linear:

$$
\begin{aligned}
s &\rightarrow \frac{2}{T}\left(\frac{1-z^{-1}}{1+z^{-1}}\right) \\
\text{so} \quad j\Omega &\rightarrow \frac{2}{T}\left(\frac{1-e^{-j\omega T}}{1+e^{-j\omega T}}\right) \\
\text{and} \quad \Omega &\rightarrow \frac{2}{T}\tan\left(\frac{\omega T}{2}\right)
\end{aligned}
$$

as in equation (4.4).

As a result, there is distortion or 'warping' of the frequency axis in going from the CT to the DT domain using the bilinear transform as previously studied, and illustrated in figure 4.4. The degree of warping increases with frequency. Strictly, this means that the bilinear transform will always produce a magnitude response in the DT realisation which is distorted in shape except when the magnitude response of the CT system being transformed is piecewise constant. If the response is constant over a range of frequencies, then warping of frequency over that range will have no effect on *shape* which will remain constant although, of course, the break frequencies will warp. The piecewise-constant condition is satisfied in the case of *ideal* low-pass, high-pass, band-pass and band-stop filters (figure 4.5).

In the case of real filters, the effects of frequency warping can be mitigated in a couple of ways. One is the obvious expedient of restricting operation to low-frequencies. The degree of warping is dependent upon the factor ωT in the argument of tan( ). So if, for instance, the sampling rate is kept high, the argument is kept low and – by the small-angle approximation $\tan(x) \approx x$ – the closeness between Ω and ω is maintained. In general, however, this may not always be a practical option.

Returning to example 2 above, the first-order Butterworth low-pass filter with cut-off frequency $\Omega_c = a$ rad.s^{-1} was found – because of frequency warping – to produce a digital filter with cut-off equal to $\frac{2}{T}\tan^{-1}(\frac{aT}{2})$ rather than a. This suggests an approach whereby we choose the CT prototype so that some significant break frequency for the analogue filter transforms down to the desired break frequency in the digital realisation. This useful technique is called *pre-warping*. We illustrate it with an example.

Example 3: synthesise a low-pass DT filter with second-order Butterworth response to have a cut-off frequency of $f_c = 50$ Hz, given that the sampling frequency is $f_s = 250$ Hz.

The frequency response of our CT prototype is given by:

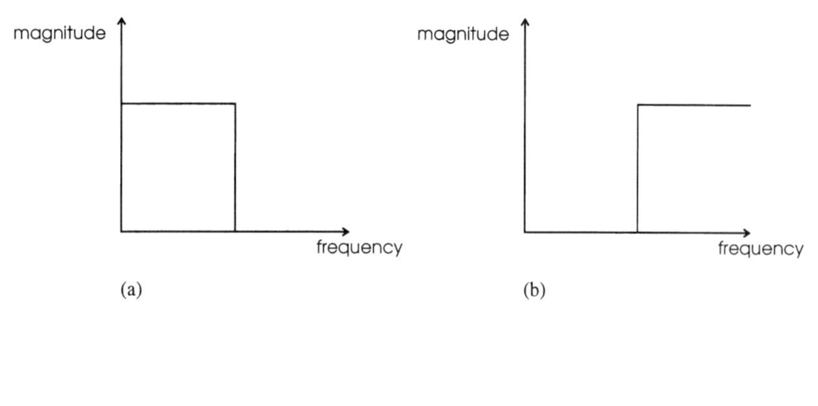

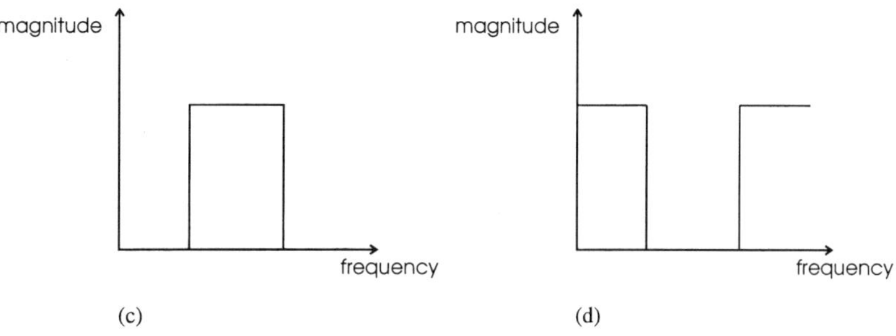

Fig. 4.5 The ideal responses of (a) low-pass, (b) high-pass, (c) band-pass and (d) band-stop filters are piecewise constant, so that they retain this property under the bilinear transform although break frequencies are warped.

$$H(s) = \frac{1}{\frac{s^2}{\Omega_c^2} + \frac{\sqrt{2}s}{\Omega_c} + 1}$$

Pre-warping to find the analogue cut-off frequency, Ω_c, from the desired digital value, ω_c:

$$\Omega_c = \frac{2}{T} \tan\left(\frac{\omega_c T}{2}\right)$$

Given that $f_c = 50$ Hz and $f_s = 250$ Hz:

$$\omega_c = 2\pi f_c = 100\pi \text{ rad.s}^{-1}$$

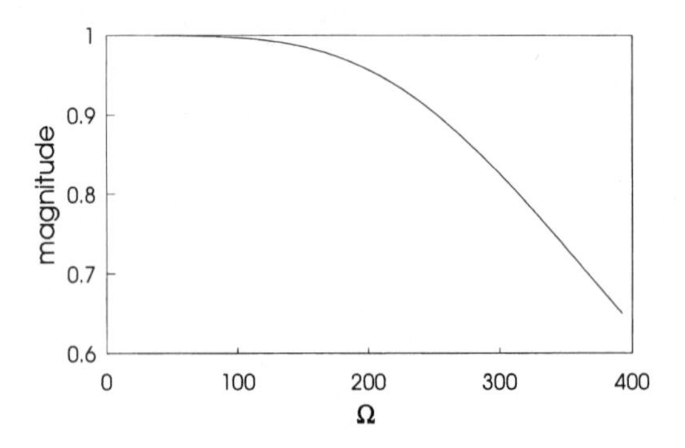

Fig. 4.6 Magnitude response of second-order, low-pass Butterworth continuous-time filter acting as prototype for the synthesis of a discrete-time version.

$$\text{and} \quad T = \tfrac{1}{250}\ \text{s} = 4\ \text{ms}$$

Hence:

$$\Omega_c = \frac{2}{4 \times 10^{-3}} \tan\left(50\pi \times 4 \times 10^{-3}\right) = 363.27\ \text{rad.s}^{-1}$$

This is the break frequency which, as a result of frequency warping, is to transform down to $\omega_c = 100\pi = 314.2$ rad.s^{-1}. Figure 4.6 shows the corresponding magnitude response of the analogue prototype, $|H(j\Omega)|$, for this cut-off.

Thus:

$$H(s) = \frac{1}{\frac{s^2}{131965} + \frac{\sqrt{2}s}{363.27} + 1} \tag{4.5}$$

Now s is to be replaced by:

$$\frac{2}{T}\left(\frac{1 - z^{-1}}{1 + z^{-1}}\right) = 500\left(\frac{1 - z^{-1}}{1 + z^{-1}}\right) \tag{4.6}$$

Substituting equation (4.6) into equation (4.5) and simplifying:

$$H(z) = \frac{1 + 2z^{-1} + z^{-2}}{4.8410 - 1.7889z^{-1} + 0.9479z^{-2}}$$

Comparing with the general rational polynomial form for a second-order DT system:

$$H(z) = \frac{b_0 + b_1 z^{-1} + b_2 z^{-2}}{1 + a_1 z^{-1} + a_2 z^{-2}}$$

gives the coefficients:

$$\begin{aligned} b_0 &= \tfrac{1}{4.8410} = & 0.2066 \\ b_1 &= \quad \ldots & 0.4131 \\ b_2 &= \quad \ldots & 0.2066 \\ a_1 &= \quad \ldots & -0.3695 \\ a_2 &= \quad \ldots & 0.1958 \end{aligned}$$

Hence, the difference equation for the second-order DT Butterworth filter is:

$$\begin{aligned} y_n &= 0.2066x_n + 0.4131x_{n-1} + 0.2066x_{n-2} \\ &\quad +0.3695y_{n-1} - 0.1958y_{n-2} \end{aligned}$$

which gives, of course, a recursive, IIR realisation.

From $H(z)$ above, this DT system has a second-order zero at $z = -1$ and poles at:

$$0.1848 \pm j0.4021 = 0.4425\angle \pm 65^0$$

as depicted in figure 4.7.

Let us use the graphical technique of the previous chapter to evaluate the magnitude of the frequency response at a few specific frequencies, as well as comparing the response of the derived DT filter with that of the CT prototype.

For a general value of ωT, there will be a general, corresponding $e^{j\omega T}$ point G on the unit circle such that:

$$M(\omega T) = k\frac{(\text{ZG})^2}{(\text{P}_1\text{G})(\text{P}_2\text{G})} \tag{4.7}$$

where k is the gain term $|\frac{b_0}{a_0}|$ $(= \frac{1}{4.8410})$, and ZG, P_1G and P_2G are the lengths of the vectors from the second-order zero and each of the two poles to the point G respectively.

When $\omega T = 0$, using some elementary trigonometry to evaluate the vector lengths in equation (4.7):

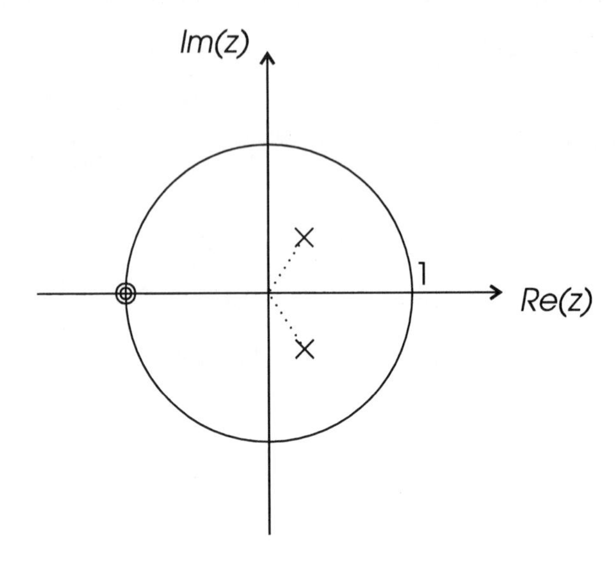

Fig. 4.7 Pole-zero diagram for DT second-order Butterworth low-pass-filter, having a second-order zero at $z = -1$ and complex-conjugate poles at $0.1848 \pm j0.4021$.

$$\begin{aligned} M(0) &= \frac{2^2}{(4.8410)(0.909)(0.909)} \\ &= 1 \qquad (= 0 \text{ dB}) \end{aligned}$$

The digital filter has unity gain at zero frequency, as can be easily checked by substituting $z = 1$ in $H(z)$. The analogue prototype also has unity zero-frequency gain. This is not surprising as there is, of course, no frequency warping at zero frequency.

At the digital cut-off frequency, $\omega_c = 314.2$ rad.s^{-1} and:

$$\begin{aligned} \omega T = \angle z &= 314.2 \times 4 \times 10^{-3} \text{ rad} \\ &= 1.2568 \text{ rad} = 72^0 \end{aligned}$$

At this point on the unit circle, again using elementary trigonometry (cosine rule):

$$M(\omega T) = \frac{(1.6180)^2}{(4.8410)(0.5634)(1.3576)}$$

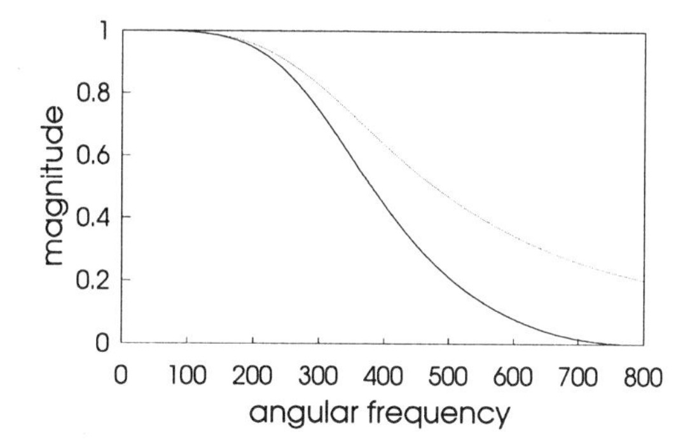

Fig. 4.8 Magnitude responses for second-order, continuous-time Butterworth low-pass filter prototype (dotted line), and the discrete-time filter derived from it using the bilinear transform (full line). For the CT filter, the response is plotted against analogue frequency, Ω. For the DT filter, the response is plotted against ωT where ω is the digital frequency and $T = 4$ ms. The (digital) cut-off frequency is $\omega_c = 314.2$ rad.s^{-1} and the folding (Nyquist) frequency is $\omega_N = 785.4$ rad.s^{-1}.

$$= \quad 0.707 \qquad (= -3 \text{ dB})$$

in accordance with the definition of the cut-off frequency as the 3 dB-down, or half-power, point.

The corresponding magnitude for the analogue prototype at $\Omega = 314.2$ rad.s^{-1} is 0.8008 (figure 4.6).

When $\omega T = 2.094$ rad $= 120^0$:

$$\begin{aligned} M(\omega T) &= \frac{1}{(4.8410)(0.8296)(1.4413)} \\ &= 0.1728 \qquad (= -15.25 \text{ dB}) \end{aligned}$$

This corresponds to a frequency of 523 rad.s^{-1}, at which the magnitude of the analogue filter is 0.4345, i.e. very considerably greater than that of the digital filter.

Figure 4.8 shows the two magnitude responses, for the CT prototype and DT derivative. Note that the DT filter has much faster roll-off as a consequence of the frequency warping. As ωT approaches π rad (corresponding to the folding frequency $\omega = \pi/T$ equals 785.4 rad.s^{-1}), so Ω approaches ∞. Notice also that, although the digital filter is a second-order system with complex-conjugate poles, there is no resonant peak in the magnitude response. This is essentially because of the second-order zero at $\omega T = \pi$.

As the vector length P_1G in equation (4.7) reduces as the $e^{j\omega T}$ point moves closer to the pole at $0.4424\angle 65^0$, so the length ZG reduces also and the two effects largely cancel. Further, the damping is quite heavy ($\zeta \approx 0.58$) since the poles are quite close to the z-plane origin.

4.1.4 Utility of the method

Use of the bilinear transform to derive a DT filter from a previously-designed CT prototype has many advantages. Principally, these are:

- the mapping function is reasonably simple.
- because of the band-limiting nature of the transformation, there is no aliasing in the frequency response (as there is with the impulse invariance method to be described).
- stable CT prototypes transform to stable DT derivatives.

There are, however, some attendant disadvantages. First, there is a warping of the digital frequency which means that break frequencies in the prototype and the DT realisation are different. Except for the (ideal) case where the analogue prototype has a piecewise-constant magnitude response, the obtained magnitude response is also distorted. The effect on break frequency can be mitigated by pre-warping the prototype's magnitude response. Second, because the bilinear transform pays no attention to either the impulse response or the phase response of the DT realisation, these are generally different to those of the CT prototype (see exercise 4.3).

4.2 IMPULSE INVARIANCE

The basis of this approach is that we aim to preserve the impulse response of the CT prototype filter when converting to a DT realisation, rather than its frequency-magnitude response as with the bilinear transform.

The unit-sample (impulse) response is thus a sampled version of the impulse response of the CT filter:

$$h_{DT}(nT) = h_{CT}(t)|_{t=nT}$$

and, unlike a design based on the bilinear transform, the time responses of the two systems are equivalent.

In section 3.10 of the previous chapter, we showed that the time response of a DT system depends principally upon the poles (rather than the zeros) of $H(z)$ and of $X(z)$.

Not surprisingly, a similar argument can be made for a CT system. Hence, the impulse response of a system is primarily determined by the poles of the system function, and we have no need explicitly to consider the zeros of the prototype $H(s)$. In what follows, therefore, we look at the system poles only. Additional insight into this issue can be gained by considering the inverse Laplace transform of $H(s) = s$ (i.e. the system function of an ideal differentiator). In this case, the inverse, $h(t)$, is the derivative of the Dirac delta $\delta(t)$. Clearly, deriving a sampled-data version of this is not a very practical proposition!

Because of the dependence of the impulse response on the system poles, the impulse response of the CT prototype is infinite and, hence, the impulse invariance method yields an IIR digital realisation.

4.2.1 Prototypes with simple poles

Consider a prototype, analogue filter with N simple (real) poles only in s:

$$H(s) = \sum_{r=1}^{N} \frac{C_r}{s + d_r} \tag{4.8}$$

From the previous chapter, by analogy between CT and DT systems, C_r is the residue at the pole:

$$C_r = H(s).(s + d_r)|_{s=d_r}$$

Taking inverse Laplace transforms:

$$\begin{aligned} h(t) &= \mathcal{L}^{-1}[H(s)] \\ &= \sum_{r=1}^{N} C_r \mathcal{L}^{-1}\left[\frac{1}{s + d_r}\right] \qquad t \geq 0 \\ &= \sum_{r=1}^{N} C_r e^{-d_r t} u(t) \end{aligned}$$

where $u(t)$ is the unit-step function ensuring causality, i.e. $h(t)$ is 0 for $t < 0$.

Hence, by the principle of impulse invariance:

$$h(nT) = \sum_{r=1}^{N} C_r e^{-d_r nT} u(nT)$$

Taking z-transforms:

$$
\begin{aligned}
H(z) &= \sum_{n=0}^{\infty} h(nT) z^{-n} \\
&= \sum_{n=0}^{\infty} \sum_{r=1}^{N} C_r e^{-d_r nT} u(nT) z^{-n}
\end{aligned}
$$

We can now delete the unit-step term, since the summation in n ensures that the argument of $u()$ is always positive, and interchange the order of summation:

$$
H(z) = \sum_{r=1}^{N} C_r \sum_{n=0}^{\infty} \left(e^{-d_r T} z^{-1}\right)^n
$$

By geometric series:

$$
\sum_{n=0}^{\infty} \left(e^{-d_r T} z^{-1}\right)^n = \frac{1}{1 - e^{-d_r T} z^{-1}}
$$

$$
\text{and} \quad H(z) = \sum_{r=1}^{N} \frac{C_r}{1 - e^{d_r T} z^{-1}} \tag{4.9}
$$

Comparing equation (4.9) with the starting point, equation (4.8), the required impulse invariance mapping for the simple pole is:

$$
\frac{1}{s + d_r} \rightarrow \frac{1}{1 - e^{-d_r T} z^{-1}} \tag{4.10}
$$

Thus, a simple pole at $-d_r$ in s maps to a real-axis pole at $e^{-d_r T}$ in z. This follows more-or-less directly from the use of the (standard) z-transform in the derivation above. Thus, if the CT prototype is stable (poles in the left-half plane of s), so too will be the derived DT filter (poles inside the unit circle of z).

4.2.2 Frequency aliasing

The use of the standard z-transform in the impulse invariance mapping indicates that there will be a problem of frequency aliasing, since this transform is not band-limited. The problem is illustrated in figure 4.9 for the case of a low-pass filter. At any frequency in the base-band $0 \leq \omega T \leq \pi$, the magnitude response is the sum of the base-band response plus components folded into this frequency region from all other bands. Clearly,

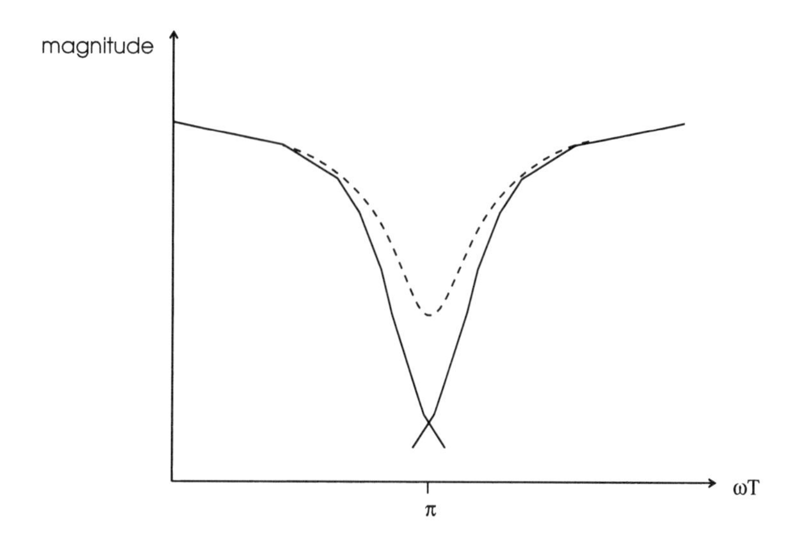

Fig. 4.9 The impulse invariance method of digital filter design leads to aliasing of the magnitude response, shown here for the case of a low-pass filter, since the transformation between continuous- and discrete-time domains is not band-limited.

this problem would be greatly exacerbated for the case of a high-pass filter, where the response does not fall off close to the folding frequency. Accordingly, the frequency response of the CT prototype filter must itself be band-limited or the aliasing problem can be severe.

Provided the prototype's response is band-limited, the problem of aliasing is usually manageable (see below), so that the frequency response of the derived DT filter can generally be made a reasonable approximation to that of the prototype. In particular, there is no frequency warping (see exercise 4.2).

4.2.3 Prototypes with complex poles

Consider a CT filter with a complex pole-pair:

$$\begin{aligned} H(s) &= \frac{s+a}{(s+a)^2+b^2} \\ &= \frac{s+a}{(s+a+jb)(s+a-jb)} \end{aligned}$$

By partial fractions:

$$H(s) = \frac{A}{s+a+jb} + \frac{B}{s+a-jb}$$

where:

$$\begin{aligned} A &= H(s).(s+a+jb)|_{s=-a-jb} \\ &= \left.\frac{s+a}{s+a-jb}\right|_{s=-a-jb} \\ &= \frac{1}{2} \end{aligned}$$

and:

$$\begin{aligned} B &= \left.\frac{s+a}{s+a+jb}\right|_{s=-a+jb} \\ &= \frac{1}{2} \end{aligned}$$

Hence:

$$H(s) = \frac{\frac{1}{2}}{s+a+jb} + \frac{\frac{1}{2}}{s+a-jb}$$

Transforming by impulse invariance:

$$\begin{aligned} H(z) &= \frac{\frac{1}{2}}{\left(1-e^{-(a+jb)T}z^{-1}\right)} + \frac{\frac{1}{2}}{\left(1-e^{-(a-jb)T}z^{-1}\right)} \\ &= \frac{1-\frac{1}{2}e^{-aT}z^{-1}\left(e^{jbT}+e^{-jbT}\right)}{1-e^{-aT}z^{-1}\left(e^{jbT}+e^{-jbT}\right)+e^{-2aT}z^{-2}} \\ &= \frac{1-e^{-aT}\cos(bT)z^{-1}}{1-2e^{-aT}\cos(bT)z^{-1}+e^{-2aT}z^{-2}} \end{aligned} \quad (4.11)$$

Let us now compare impulse responses of the prototype and derived filters for this example. For the CT prototype:

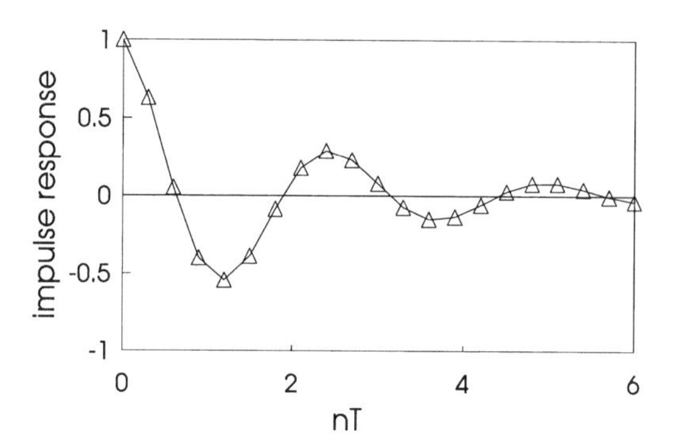

Fig. 4.10 Impulse response of the DT system derived from a CT prototype, $h_{DT}(nT)$, is a sampled version of the impulse response of the prototype, $h_{CT}(t)$, shown here for the values $a = 0.5, b = 2.5$ and $T = 1$ s.

$$
\begin{aligned}
h_{CT}(t) &= \mathcal{L}^{-1}\left[\frac{s+a}{(s+a)^2+b^2}\right] \\
&= e^{-at}\cos(bt)
\end{aligned}
$$

Taking equation(4.11) and recognising that each term is a geometric series:

$$
\begin{aligned}
H(z) &= \frac{1}{2}\left\{\sum_{n=0}^{\infty}\left(e^{-(a+jb)T}z^{-1}\right)^n + \sum_{n=0}^{\infty}\left(e^{-(a-jb)T}z^{-1}\right)^n\right\} \\
\text{so that } h_{DT}(nT) &= \frac{1}{2}\left\{e^{-(a+jb)nT} + e^{(a-jb)nT}\right\} \\
\text{and } h(n) &= e^{-anT}\cos(bnT) \qquad (4.12)
\end{aligned}
$$

Hence, $h(n)$ is a sampled version of $h_{CT}(t)$ as required. In general, therefore, the impulse invariance method will yield an IIR filter.

Figure 4.10 shows the impulse response for the case $a = 0.5, b = 2.5$ and $T = 1$ s.

As an illustration, we now compare frequency-magnitude responses for the CT and DT systems for the specific values $a = \frac{1}{2}$ rad.s^{-1}, $b = \frac{1}{2}$ rad.s^{-1} and $T = 2$ s.

$$
H(s) = \frac{s+\frac{1}{2}}{\left(s+\frac{1}{2}+j\frac{1}{2}\right)\left(s+\frac{1}{2}-j\frac{1}{2}\right)}
$$

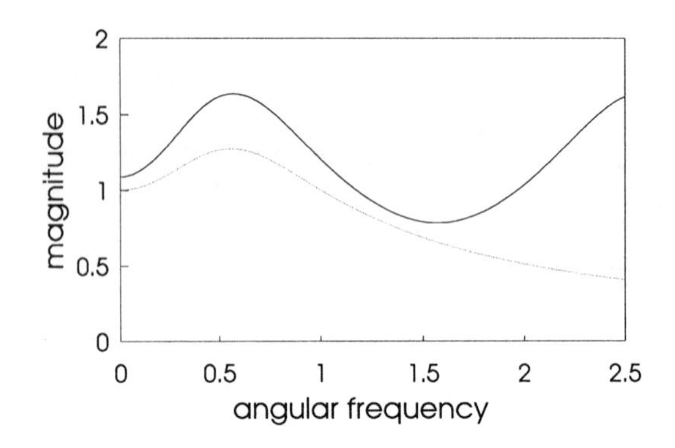

Fig. 4.11 Magnitude responses of second-order CT filter (dotted line) and the DT filter derived from it by impulse invariance (full line). For the DT filter, the response is plotted against ωT where ω is the digital frequency and $T = 2$ ms. The folding (Nyquist) frequency is $\omega_N = \frac{\pi}{2}$ rad.s^{-1}.

There is a zero at $s = -\frac{1}{2}$ and a pole pair at $s = -\frac{1}{2} \pm j\frac{1}{2}$.
From equation (4.11):

$$
\begin{aligned}
H(z) &= \frac{1 - e^{-1}\cos(1)z^{-1}}{1 - 2e^{-1}\cos(1)z^{-1} + e^{-2}z^{-2}} \\
&= \frac{z^2 - 0.199z}{z^2 - 0.398z + 0.135}
\end{aligned}
$$

So the DT filter has z-plane zeros at 0 and 0.199, and poles at $0.199 \pm j0.310$.

While the s-plane poles ($a \pm jb$) map directly to z-plane poles ($e^{-aT}e^{-jbT}$), the zeros of $H(z)$ are a function of the s-plane poles and the residues. The direct mapping of the poles means that stability of the CT prototype will be preserved.

Figure 4.11 shows the magnitude responses for the CT prototype and CT derivative. The general shape in the interval $(0, \frac{\pi}{2})$ is similar except:

- the absolute magnitudes are different;
- the roll-off of the DT filter is much slower than that of the CT filter because of aliasing;
- the resonant peaks of these second-order systems do not occur at exactly the same frequency, again because of aliasing.

We illustrate the latter point by evaluating the resonant or 'peak' frequencies, Ω_p and ω_p, for the CT and DT filters respectively.

For the CT system:

$$|H(j\Omega)| = \sqrt{\frac{\frac{1}{4} + \Omega^2}{\frac{1}{4} + \Omega^4}}$$

The resonant peak occurs for:

$$\begin{aligned} \frac{d|H(j\Omega)|^2}{d\Omega} &= 0 \\ \text{and} \quad 2\Omega_p^4 + \Omega_p^2 - \frac{1}{2} &= 0 \end{aligned}$$

or $\Omega_p = 0.556$ rad.s^{-1}.

For the DT system, the magnitude response $|H(e^{j\omega T})|$ is given by:

$$\sqrt{\frac{(1 - 0.199\cos\omega T)^2 + \sin^2\omega T}{(1 - 0.398\cos\omega T + 0.135\cos 2\omega T)^2 + (0.398\sin\omega T - 0.135\sin 2\omega T)^2}}$$

This is not at all easy to differentiate! Using a spreadsheet, $\omega_p T$ is found to be 0.569 rad.s^{-1} (c.f. $\Omega_p = 0.556$ rad.s^{-1}).

4.2.4 Summary of design procedure

The procedure of designing a DT filter from a CT prototype using the method of impulse invariance can be summarised as follows. First, we expand $H(s)$ for the prototype into partial fractions. Then, we apply the impulse invariance transformation, equation (4.10) after which we correct for differences in absolute magnitude, if necessary. We can then examine the effects of aliasing on the magnitude and phase responses. If the aliasing effects are outside of acceptable limits, we modify $H(s)$ and try again.

Hence, design based on impulse invariance is an iterative, trial-and-error procedure.

4.3 SUMMARY

Unfortunately, there are no general methods available to synthesise a discrete-time system (i.e. to find an appropriate system function $H(z)$) from arbitrary, desired magnitude and phase responses. This is not true, however, of continuous-time systems where methods do exist to find an $H(s)$ corresponding to desired frequency responses. In this chapter, we have studied two very useful methods for deriving discrete-time ('digital') filters from continuous-time ('analogue') prototypes. Both yield filters with infinite impulse responses so that stability is an issue.

The first of these – the bilinear-transform method – is two-step in essence. The first step is a band-limiting process, based on the tanh function, which maps the entire s-plane to a strip bounded by $\pm j\pi/T$ in some intermediate domain, w. The second step employs the 'standard' z-transform, $w \rightarrow z = e^{wT}$, to map to the bilinear z-plane. Because the transform is inherently band-limited, problems of frequency aliasing are avoided. There is, however, a 'warping' of the frequency axis in the DT domain, so that the break frequencies of the CT prototype and digital-filter realisation differ. Further, unless the prototype frequency response is piecewise constant, the shape of the DT response is distorted. The problem of different break frequencies can be addressed by *pre-warping* the response of the CT prototype so that, after transformation, the desired DT break frequency results. Because the second step is based on the 'standard' z-transform, the left-half s-plane maps to the inside of the (bilinear) z-plane unit circle, and the bilinear-transform method yields a stable digital-filter realisation provided the CT prototype is stable. The transform always produces a DT realisation with poles (i.e. a recursive realisation) even when the prototype has only zeros. Hence, the digital filter produced has an infinite impulse response. Finally, the method attempts only to retain the magnitude response of the prototype: neither the impulse response nor the phase response are preserved.

The second method studied was impulse invariance which, as the name suggests, attempts to preserve the impulse response of the CT prototype. Because the impulse response depends upon the poles of the system function, the DT realisation is an IIR filter. Because it is based on use of the standard z-transform, the method also yields a stable digital filter provided the analogue prototype is stable. While there is no frequency warping, aliasing does occur and can be serious in particular cases. An especially severe limitation is that the method is only really appropriate for low-pass (and some band-pass) designs. The problem of aliasing, together with the concentration on preserving the impulse response, means that the derived filter (even if low-pass) can have magnitude and phase responses which are a poor match to the prototype's. In this case, an iterative adjustment of $H(s)$ can be tried.

4.4 EXERCISES

4.1 Show that the bilinear transform can be written in the form:

$$Asz + Bs + Cz + D = 0$$

and evaluate the constants A, B, C and D. (Incidentally, this is the reason for the name *bilinear*: this equation is linear in s, with z considered constant, and linear in z, with s considered constant.)

4.2 When using the bilinear transform, we have to take explicit measures to deal with frequency warping. Show that there is no frequency warping with the 'standard' z-transform ($z = e^{sT}$).

4.3 The bilinear transform aims to preserve the magnitude response of a continuous-time prototype filter when deriving a discrete-time realisation. However, no consideration is given to either the impulse or phase responses. Evaluate these for both CT and DT versions of the first-order Butterworth low-pass filter of example 2 and and compare them.

4.4 Consider the CT filter with system function $H(s) = b/(s+a)$, which is also described by the differential equation:

$$\frac{dy(t)}{dt} + ay(t) = bx(t)$$

Show by integrating this equation and converting it to DT form (making the substitution $t = nT$ and taking z-transforms) that use of the bilinear transform corresponds to approximating the CT system by using the trapezoidal numerical-integration formula.

4.5 Repeat the low-pass filter design of example 3 for the lower sampling frequency of $f_s = 150$ Hz. In general, what is the effect of a change of sampling rate on a digital filter design based on the bilinear transform?

4.6 Design a high-pass, second-order Butterworth digital filter with cut-off frequency $f_c = 100$ Hz and sampling frequency $f_s = 250$ Hz. Hence, using the result of example 3, obtain a band-pass filter with pass-band 50–100 Hz.

4.7 For the low-pass digital filter design of example 3, the cut-off frequency obtained at $\omega_c T = 1.2568$ rad $= 72^0$, while the angle θ associated with the poles at $r\angle \pm \theta$ was 65^0. Obtain a general expression relating $\omega_c T$ and θ.

4.8 Repeat the design of the discrete differentiator (example 1) with $T = 1$ second, using the impulse invariance method in place of the bilinear transform. Compare the frequency responses (magnitude and phase) of the two digital realisations.

4.9 Repeat the low-pass, first-order Butterworth design of example 2 using the impulse invariance method in place of the bilinear transform, for the same values of a and T (10π rad.s^{-1} and 25 ms respectively). Compare the frequency responses (magnitude and phase) of the two digital realisations. How will the design change if the sampling period is (a) doubled to 50 ms and (b) halved to 12.5 ms?

4.10 Repeat the low-pass, second-order Butterworth design of example 3 using the impulse invariance method in place of the bilinear transform, for the same values of cut-off frequency ($f_c = 50$ Hz) and sampling frequency ($f_s = 250$ Hz). Again, compare the frequency responses (magnitude and phase) of the two digital realisations.

4.11 Figure 4.11 illustrated the fact that the absolute magnitudes of the analogue prototype and digital filter responses differ. Modify the digital filter response so that the zero-frequency gains coincide (i.e. both unity). What is the percentage error in peak response in this case? Now modify the digital filter response so that the peak responses coincide. What is then the percentage error in zero-frequency response?

4.12 Since the unit-sample response of the DT filter derived from a CT prototype by impulse invariance, $h_{DT}(nT)$, is a sampled version of the prototype's impulse response $h_{CT}(t)$, a question which arises is this. Is it possible for $h_{DT}(nT)$ to be under-sampled? Consider this question with reference to equation (4.12).
(You may recognise this question as related to exercise 1.3.)

5

The discrete Fourier transform

In this chapter, we are centrally concerned with the Fourier representation of finite-length sequences, i.e. discrete-time signals, which are sampled versions of continuous-time signals, with window length N. That is, we have some CT signal $x(t)$, with spectrum $X(j\omega)$, which is sampled to produce $x^*(t)$ or $x(n)$ which represents $x(t)$ over some finite interval corresponding to $0 \leq n \leq (N-1)$. In general, because the sequence values result from sampling real signals, they will be real also. Whatever representation we choose for the spectrum of $x(n)$, it should have a direct relation to $X(j\omega)$. We shall see that a variety of spectral representations is possible but one of them – the discrete Fourier transform (DFT) – offers particular advantages.

The Fourier series and the Fourier transform can be used to represent CT signals that are either periodic or aperiodic respectively, and it is possible to develop discrete-time versions of both of these. A difficulty already encountered (chapter 3) with the direct application of the Fourier transform to a DT signal is that the resulting representation is a continuous function of frequency. Hence, it is unsuitable for digital processing. The obvious problem with the discrete Fourier series is that an arbitrary DT signal, in the form of a sequence of finite length N, will not generally be periodic. We will find that the DFT offers a way out of this difficulty by treating the signal as if it were periodic in time with period equal to the length N of the analysis window.

The importance of the DFT to the study of discrete-time signals and systems could hardly be over-stated. We have previously seen how a Fourier representation of a system's impulse response (i.e. the frequency response) can aid understanding of system properties. Accordingly, in this chapter we will meet again several properties that we have already seen. However, the treatment here is considerably more general, and in greater depth. In the next chapter, we will have occasion to use the DFT in the design of FIR digital filters. Further, as we already know for the CT case, many signals can be more easily understood from their spectral, frequency-domain representation than from the time-domain waveform. In particular, signal changes that extend widely over time tend to be compact in frequency. For instance, a CT sinusoid of frequency ω_0 and infinite extent in time can be represented using the Fourier series with spectral components at $\pm\omega_0$ only, while many complex periodic signals can be represented as a fundamental component plus a small number of harmonics of the fundamental. This is especially useful in the study of many real-world signals – such as speech, and vibrations in structures driven by a rotating engine – which are *quasi*-periodic either in part or in whole. Finally, many signal-processing operations can be carried out much more efficiently in the frequency-domain than in the time-domain. A good example of this is convolution, which is (as we have already seen in our study of the z-transform) a rather complex operation in the time-domain but can be effected as simple multiplication in the frequency (z-)domain. The existence of a class of efficient algorithms (fast Fourier transforms or FFTs) for computing the DFT greatly increases the practicality and attractiveness of frequency-domain processing.

Before defining the DFT, we re-examine the result of applying the Fourier transform to a sampled signal.

5.1 Spectrum of a sampled signal

In section 3.3, we used the rectangular rule to approximate the Fourier transform of a sequence as:

$$X(j\omega) \approx \left(\sum_{n=0}^{\infty} x(n) e^{-j\omega nT} \right) T \, \frac{\sin(\omega T/2)}{(\omega T/2)} \tag{5.1}$$

Since the sinc term is independent of n, and acts as a scaling multiplier for any given value of ω, it suffices to take the bracketed term as our Fourier representation, $X^*(j\omega)$. We emphasised that, as a result of the $e^{-j\omega nT}$ term, this spectral representation is periodic in frequency with period $2\pi/T$.

For present purposes, there are two problems with $X^*(j\omega)$ as a frequency-domain representation. First, it assumes that the sampled data sequence $x(n)$ is infinite in extent. Second, it is a continuous function of frequency and, as such, is inappropriate for digital processing. We will shortly define the DFT so as to overcome these objections. Before so doing, however, we examine a slightly different spectral representation of a band-limited sampled signal, $X^*(j\omega)$. Our main purpose here is to gain additional understanding of the relation between $X^*(j\omega)$ and $X(j\omega)$, and especially the periodicity properties of $X^*(j\omega)$ since these will also apply to the DFT. Unlike equation (5.1), this new representation will not involve any approximation.

Consider a real CT signal $x(t)$ band-limited to a frequency $f_{max} = B$ with Fourier spectrum $X(j\omega)$. The sampled version of $x(t)$ can be denoted $x^*(t)$ where, from equations (1.8) and (1.7):

$$x^*(t) = x(t)c_T(t) = x(t)\sum_{n=-\infty}^{\infty}\delta(t-nT)$$

The comb function $c_T(t)$ consists of a train of Dirac pulses separated in time by T, and so it is clearly periodic with period T. Its Fourier series representation is:

$$c_T(t) = \sum_{n=-\infty}^{\infty} c_n e^{jn\omega_s t}$$

with complex coefficients $$c_n = \frac{1}{T}\int_0^T c_T(t)e^{-jn\omega_s t}dt$$

and $$\omega_s = \frac{2\pi}{T}$$

Since the only non-zero value of $c_T(t)$ in the interval $0 \le t < T$ is $\delta(t)|_{t=0}$ then, by the sifting property, $c_n = (1/T)e^0 = 1/T$ for all n. Hence:

$$x^*(t) = \frac{1}{T}\sum_{n=-\infty}^{\infty} x(t)e^{jn\omega_s t} \tag{5.2}$$

Now the FT of $x^*(t)$ is:

$$X^*(j\omega) = \int_{-\infty}^{\infty} x^*(t)e^{-j\omega t}dt$$

Substituting from equation (5.2):

$$X^*(j\omega) = \frac{1}{T}\int_{-\infty}^{\infty}\left(\sum_{n=-\infty}^{\infty} x(t)e^{jn\omega_s t}\right)e^{-j\omega t}dt$$

Interchanging the order of summation and integration:

$$X^*(j\omega) = \frac{1}{T}\sum_{n=-\infty}^{\infty}\left(\int_{-\infty}^{\infty} x(t)e^{-j(\omega-n\omega_s)t}dt\right)$$

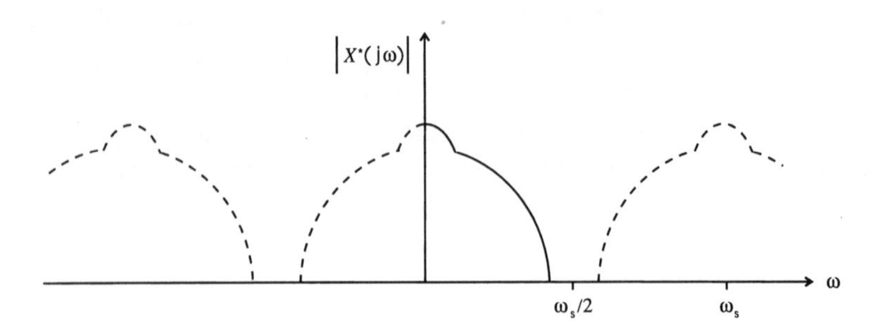

Fig. 5.1 The spectrum $X^*(j\omega)$ of a sampled signal $x^*(t)$ is periodic in frequency with period ω_s. In the baseband, $-\omega_s/2 \leq \omega \leq \omega_s/2$, it is proportional to the spectrum $X(j\omega)$ of the original CT signal $x(t)$, as depicted here for the magnitude spectrum $|X^*(j\omega)|$. The constant of proportionality is $1/T$.

But, from the definition of the FT, the bracketed integral is $X(j\omega - jn\omega_s)$, so that:

$$X^*(j\omega) = \frac{1}{T} \sum_{n=-\infty}^{\infty} X(j\omega - jn\omega_s) \tag{5.3}$$

This quantity is, of course, complex. As we did in the case of the frequency response of a DT system (chapter 3), we can define a magnitude and a phase spectrum as $M^*(j\omega) = |X^*(j\omega)|$ and $\Phi^*(j\omega) = \angle M^*(j\omega)$ respectively.

Since $x(t)$ is band-limited to frequency B, $X(j\omega - jn\omega_s)$ in equation (5.3) is only non-zero when its argument is in the range $(-B, +B)$. When $n = 0$, $X(j\omega - jn\omega_s)$ is simply equal to $X(j\omega)$, and the contribution to $X^*(j\omega)$ is the baseband spectrum of $x(t)$ scaled by $1/T$. When $n = 1$, this baseband contribution is shifted by $-\omega_s$ in frequency, and when $n = -1$, it is shifted by $+\omega_s$. Because of the band-limiting, none of these contributions (extending to $\pm\infty$) overlap.

The overall relation between $X^*(j\omega)$ and $X(j\omega)$ is, therefore, as depicted in figure 5.1 where we show just the magnitude spectrum for convenience. (Note that we have shown the magnitude spectrum as an even function of frequency: we will look more closely at the symmetry properties of the discrete Fourier transform below in section 5.5.6. They are very similar to those $X^*(j\omega)$.) This figure portrays the important results that, for a real CT signal $x(t)$ band-limited to frequency B and properly sampled with $f_s \geq 2B$ to yield a DT signal $x^*(t)$:

- the Fourier transform, $X^*(j\omega)$, of $x^*(t)$ is periodic in frequency with a period of $\omega_s = 2\pi/T$;

- within any one period, this transform is proportional in magnitude to the FT of the CT signal, $X(j\omega)$, scaled by $1/T$.

Because of the way it is arrived at from the Fourier transform, the discrete Fourier transform has very similar properties as we will soon find.

It can also be seen from figure 5.1 that the reconstitution of the waveshape of $x(t)$ from $x^*(t)$ is, in principle, accomplished by passing the sampled signal through an ideal low-pass filter with cut-off frequency B (or, equivalently, f_s) to leave the baseband contribution only.

If the signal is not properly sampled, i.e. $f_{\max} > f_s/2$, it is apparent that the different contributions for adjacent values of n will overlap rather than remaining separate. This is the now-familiar phenomenon of aliasing, whereby a high-frequency component above the folding or Nyquist point ($f_N = f_s/2$) cannot be distinguished from one appearing in the baseband. For example, if $f_s = 200$ Hz ($f_N = 100$ Hz) then a 110 Hz component in $x(t)$ (i.e. 10 Hz above f_N) will appear at 90 Hz (i.e. 10 Hz below f_N) in $X^*(j\omega)$. When reconstituting $x(t)$ from this spectrum by low-pass filtering, the 110 Hz component will appear as a lower-frequency alias at 90 Hz.

While the $X^*(j\omega)$ representation emphasises some of the properties of Fourier spectra of sampled signals, it is generally unsuitable for practical applications in digital processing. We require a spectral representation which retains the essential properties of $X(j\omega)$ in the baseband yet is discrete in frequency and can be applied to finite-length sequences. This representation is the discrete Fourier transform.

5.2 DEFINITION OF THE DFT

From earlier (chapter 3 and the previous section), we have the following FT representation of a sampled signal:

$$X^*(j\omega) = \sum_{n=0}^{\infty} x(n)e^{-j\omega nT}$$

However, this form assumes an infinite-length sequence and, further, $X^*(j\omega)$ is a continuous function of frequency. In defining the DFT, we seek to overcome both these objections.

To make the transform a discrete function of frequency, we first note that the Fourier spectrum of a sampled-data sequence is periodic with period $\omega = 2\pi/T$. If our transform is to be computationally efficient, it suffices to represent a single period of the spectrum. So, making the frequency spacing $2\pi/NT$ we then have a length-N transform. Hence, we replace the discrete frequency variable ω with $2\pi m/NT$, where m is an index of (discrete) frequency, to give:

$$X(m) = \sum_{n=0}^{N-1} x(n) e^{-2\pi jmn/N} \quad (5.4)$$

This, by definition, is the discrete Fourier transform. It specifies how to analyse $x(n)$ into a set of complex exponentials which are discrete in both frequency and time. We write:

$$X(m) = \mathcal{D}[x(n)]$$

Because of the way the DFT is defined from $X^*(j\omega)$, it shares analogous properties to those examined in the previous section (modified to take account of its discrete-in-frequency nature). That is, it is periodic in (discrete) frequency with period $n = N$. Most importantly, the baseband spectrum is a sampled and scaled version of the spectrum $X(j\omega)$ of the 'underlying' CT signal $x(t)$.

To understand this definition, let us explore the relation between equation (5.4) and the Fourier series of the sampled signal $x^*(t)$ found by sampling the band-limited CT signal $x(t)$. To find the Fourier series, we must assume that $x(t)$ is periodic in time with period $P = 1/f_0$. Its FS expansion is:

$$\begin{aligned} x(t) &= \sum_{m=-\infty}^{\infty} c_m e^{jm\omega_0 t} \\ \text{with} \quad c_m &= \frac{1}{P}\int_0^P x(t) e^{-jm\omega_0 t} \\ \text{and} \quad \omega_0 &= \frac{2\pi}{P} \end{aligned}$$

Since $x(t)$ is band-limited to frequency B, there are only a finite number of c_m's, such that:

$$c_m = 0 \text{ for } |m| > M$$

where M is some positive integer.

Hence:

$$B = Mf_0 = \frac{M}{P}$$

According to the Nyquist criterion, we require $f_s = 1/T \geq 2M/P$ in order to sample $x(t)$ properly. Let us take the equality:

$$P = 2MT$$

This not only ensures that there are an integer number of samples in the period P but also that our representation is maximally efficient, i.e. using the theoretical minimum number of coefficients. In fact, since $f_s = 2M/P$, there will be $2M$ samples per period and we can represent a single period of the sampled signal as:

$$x_P^*(t) = \sum_{n=0}^{2M-1} x(n)\delta(t - nT)$$

But $x^*(t)$ will also be periodic if $x(t)$ is. Thus, we also have the Fourier series representation:

$$\begin{aligned}
x^*(t) &= \sum_{m=-\infty}^{\infty} c'_m e^{jm\omega_0 t} \\
&= \sum_{m=-M}^{M} c'_m e^{jm\omega_0 t} \\
\text{with} \quad c'_m &= \frac{1}{P}\int_0^P x^*(t)e^{-jm\omega_0 t}\,dt
\end{aligned}$$

Since the integration is over a single period:

$$\begin{aligned}
c'_m &= \frac{1}{P}\int_0^P x_P^*(t)e^{-jm\omega_0 t}\,dt \\
&= \frac{1}{P}\int_0^P \sum_{n=0}^{2M-1} x(n)\delta(t - nT)e^{-jm\omega_0 t}\,dt \\
&= \frac{1}{P}\sum_{n=0}^{2M-1} x(n)\int_0^P \delta(t - nT)e^{-jm\omega_0 t}\,dt \\
&= \frac{1}{P}\sum_{n=0}^{2M-1} x(n)e^{-j\omega_0 mnT} \quad \text{(by the sifting property with } t = nT\text{)}
\end{aligned}$$

Now:

$$\omega_0 = \frac{2\pi}{P} = \frac{\pi}{MT}$$

Hence:

$$c'_m = \frac{1}{P}\sum_{n=0}^{2M-1} x(n)e^{-j\pi mn/M} \tag{5.5}$$

Comparing equation (5.5) and the DFT defining equation (5.4), they are identical for $2M = N$, apart from the $1/P$ multiplier. Thus, the DFT of a sequence $x(n)$ of finite length N can be interpreted as the FS coefficients of a single period of $x^*(t)$.

This suggests two possible ways of understanding the DFT:

- If the sequence $x(n)$ is truly periodic, then the $X(m)$ values obtained from the DFT are proportional to the coefficients of the discrete Fourier series. The constant of proportionality is $1/P = 1/2MT = 1/NT$.

- If $x(n)$ is actually aperiodic, as will most often be the case, then P is an abstraction. We can, however, retain it as a computational convenience. How then do we interpret P? From above, we have that $P = NT$. But a DT signal (aperiodic) represented by N sample values spaced by the sampling period T in the time-domain will obviously have duration NT seconds. Hence, we can consider the finite-duration DT signal to be periodic in time, with a period set by the length N of the analysis window. To be consistent, then, we must also consider the signal to repeat indefinitely outside the analysis window, rather than being zero, although within the window the two representations are identical. The consequences of this assumed periodicity are of some practical importance.

Whichever interpretation is the most appropriate, the baseband spectrum is in both cases a sampled and scaled version of $X(j\omega)$. It is also worth reiterating that the number (N) of discrete values used to represent the DFT is the theoretical minimum according the Nyquist sampling theorem. Hence, the DFT gives a maximally efficient frequency-domain representation of a DT signal.

In section 3.3, we considered the relation between the z-transform and the Fourier transform of a sampled-data sequence. The ZT, $X(z)$, is continuous in the complex variable $z = e^{sT} = e^{(\sigma+j\omega)T}$ and describes an infinite (right-sided) sequence. The FT representation developed in chapter 3 evaluates the z-transform around the unit circle, i.e. for $\sigma = 0$. Hence, it is also a continuous function of frequency and describes an infinite (right-sided) sequence. The description is periodic in frequency: any one period is, of course, sufficient to specify the sequence fully.

The DFT, however, is a discrete function of frequency. It is this essential property which makes it suitable for the computer processing of signals. Thus, the DFT evaluates the z-transform only at N discrete points – equally spaced – on the unit circle.

5.3 ILLUSTRATIVE EXAMPLES

In this section, we illustrate the use the DFT by finding the magnitude and phase spectra of some representative DT signals.

Example 1: Find the discrete Fourier transform (magnitude and phase) of the Kronecker delta function $x(n) = \delta_0$.

Substituting into the DFT defining equation (5.4):

$$X(m) = \mathcal{D}[\delta_0] = \sum_{n=0}^{N-1} \delta_0 e^{-2\pi jmn/N}$$

From the sifting property:

$$\begin{aligned} X(m) &= e^{-2\pi jmn/N}\Big|_{n=0} \\ &= 1 \quad \text{for all } m \end{aligned}$$

Hence, the spectral representation of the Kronecker delta function consists of equal (unit) amplitude values at all frequencies, reflecting the wideband nature of the function. Since all $X(m)$'s are real, the magnitude spectrum is identically equal to 1 and the phase spectrum is identically zero.

Example 2: Find the DFT magnitude and phase spectra of the sequence $x(n) = \cos(2\pi n/N)$.

Expressing $x(n)$ in complex-exponential form:

$$\cos(2\pi n/N) = \frac{1}{2}\left(e^{2\pi jn/N} + e^{-2\pi jn/N}\right)$$

and substituting into the DFT defining equation:

$$\begin{aligned} X(m) &= \frac{1}{2}\sum_{n=0}^{N-1}\left(e^{2\pi jn/N} + e^{-2\pi jn/N}\right)e^{-2\pi jmn/N} \\ &= \frac{1}{2}\sum_{n=0}^{N-1}\left(e^{\frac{2\pi jn}{N}(1-m)} + e^{-\frac{2\pi jn}{N}(1+m)}\right) \\ &= \frac{1}{2}\sum_{n=0}^{N-1}\left(e^{\frac{2\pi j}{N}(1-m)}\right)^n + \frac{1}{2}\sum_{n=0}^{N-1}\left(e^{-\frac{2\pi j}{N}(1+m)}\right)^n \end{aligned}$$

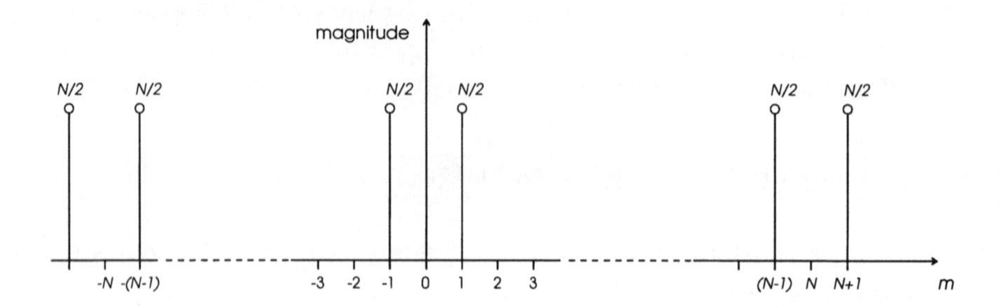

Fig. 5.2 The DFT spectrum of $\cos(2\pi n/N)$ is purely real, consisting of frequency components $N/2$ at $m = 0$ and $m = (N-1)$ in the band $0 \le m \le (N-1)$. The spectrum is periodic in the frequency index m with period $m_0 = N$.

But by the summation formula for a truncated geometric series (exercise 2.4):

$$\sum_{n=0}^{N-1} q^n = \begin{cases} N & q = 1 \\ \frac{1-q^N}{1-q} & q \ne 1 \end{cases}$$

so that:

$$\sum_{n=0}^{N-1} \left(e^{\frac{2\pi j}{N}(1-m)}\right)^n = \begin{cases} N & (1-m) = 0, \pm N, \pm 2N, \ldots \\ 0 & \text{otherwise} \end{cases}$$

Similarly:

$$\sum_{n=0}^{N-1} \left(e^{\frac{-2\pi j}{N}(1+m)}\right)^n = \begin{cases} N & (1+m) = 0, \pm N, \pm 2N, \ldots \\ 0 & \text{otherwise} \end{cases}$$

Combining these two results:

$$X(m) = \begin{cases} N/2 & m = \pm 1, \pm(N-1), \pm(N+1), \\ & \qquad \pm(2N-1), \pm(2N+1)\ldots \\ 0 & \text{otherwise} \end{cases}$$

The magnitude spectrum is depicted in figure 5.2. Since $N/2$ is purely real, the phase spectrum is identically zero. Several important points can be made about this magnitude spectrum. First of all, it is periodic in the frequency index m with period

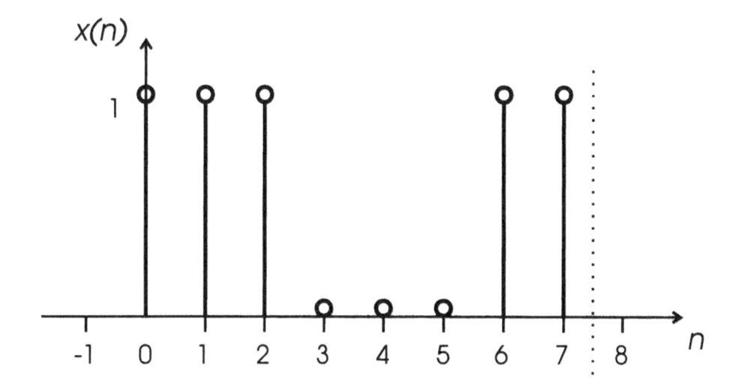

Fig. 5.3 Rectangular pulse $x(n)$ with period $n_0 = 8$.

$m_0 = N$. As expected for a pure cosine wave, most of the DFT coefficients are zero. This is a consequence of the orthogonality between $\cos(2\pi n/N)$ and the set of sine and cosine basis functions defined by $e^{-2\pi jmn/N}$. The spectrum is an even function of frequency. Also, there is even symmetry in each width-N band about the points $(\pm N/2, \pm 3N/2, \ldots)$. The point $m = N/2$ corresponds to the folding frequency. In the band $0 \leq m \leq (N-1)$, the 'power' of the signal is split between the two non-zero components at $m = 1$ and $m = (N-1)$. The fact that the spectral components appear at these particular values of m results from there being precisely one period of the signal in the length-N analysis window. That is, the index $m = 1$ corresponds to the basic frequency of the cosine wave, $\omega T = 2\pi/N$.

Example 3: Find the DFT (magnitude and phase) of the rectangular pulse with period $n_0 = 8$:

$$x(n) = \begin{cases} 1 & n = 0, 1, 2, 6, 7 \\ 0 & n = 3, 4, 5 \end{cases}$$

as depicted in figure 5.3.

From the DFT defining equation with $N = 8$:

$$\begin{aligned} X(m) &= \sum_{n=0}^{7} x(n) e^{-2\pi jmn/8} \\ &= \sum_{n=0}^{2} e^{-\pi jmn/4} + \sum_{n=6}^{7} e^{-\pi jmn/4} \end{aligned}$$

Since, for fixed m, $e^{-\pi jmn/4}$ is periodic in n with period $n_0 = 8$:

$$e^{-\pi jmn/4} = e^{-\pi jm(n-n_0)/4} = e^{-\pi jm(n-8)/4}$$

and:

$$X(m) = \sum_{n=-2}^{2} e^{-\pi jmn/4}$$

By a slightly different version of the summation formula for a finite geometric series in which the lower limit of the summation is non-zero (and may be negative):

$$\sum_{n=M}^{N-1} q^n = \begin{cases} N - M & q = 1 \\ \frac{q^M - q^N}{1-q} & q \neq 1 \end{cases} \tag{5.6}$$

so that, with $M = -2$ and $N = 3$:

$$X(m) = \begin{cases} 5 & m = 0, \pm 8, \ldots \\ \frac{e^{2\pi jm/4} - e^{-3\pi jm/4}}{1-e^{-\pi jm/4}} & \text{otherwise} \end{cases}$$

So, except for $m = 0, \pm 8, \ldots$:

$$\begin{aligned} X(m) &= \frac{e^{\pi jm/8}}{e^{\pi jm/8}} \left(\frac{e^{2\pi jm/4} - e^{-3\pi jm/4}}{1 - e^{-\pi jm/4}} \right) \\ &= \frac{e^{5\pi jm/8} - e^{-5\pi jm/8}}{e^{\pi jm/8} - e^{-\pi jm/8}} \\ &= \frac{\sin(5\pi m/8)}{\sin(\pi m/8)} \end{aligned}$$

Like the sinc function that we saw in chapter 3, this kind of $\sin(kx)/\sin(x)$ function arises frequently in signal processing. Again, this reflects the common occurrence of rectangular pulses and windows in both time and frequency domains. We will meet it again several times. It is sometimes known as the *Dirichlet function*. There is a minor problem in evaluating it for $x = 0$, since both numerator and denominator then equal zero. This difficulty is easily overcome by use of the small angle formula, $\sin(x) \approx x$ for small x. Hence:

$$\left.\frac{\sin(kx)}{\sin(x)}\right|_{x=0} = k$$

and
$$\frac{\sin(5\pi m/8)}{\sin(\pi m/8)} = 5 \quad \text{for } m = 0, \pm 8, \ldots$$

Therefore, the DFT of the rectangular pulse is:

$$X(m) = \frac{\sin(5\pi m/8)}{\sin(\pi m/8)} \qquad \text{for all } m$$

and the magnitude spectrum is:

$$|X(m)| = \left|\frac{\sin(5\pi m/8)}{\sin(\pi m/8)}\right|$$

Again, $X(m)$ is purely real and the phase spectrum is identically zero.

Example 4: Find the DFT (magnitude and phase) of the rectangular pulse in example 3 left-shifted by one sample:

$$p(n) = x(n+1) = \begin{cases} 1 & n = 0, 1, 5, 6, 7 \\ 0 & n = 2, 3, 4 \end{cases}$$

Proceeding as before:

$$X(m) = \sum_{n=0}^{1} e^{-\pi jmn/4} + \sum_{n=5}^{7} e^{-\pi jmn/4}$$

Again, using the periodicity of $e^{-\pi jmn/4}$:

$$X(m) = \sum_{n=-3}^{1} e^{-\pi jmn/4}$$

so that, from equation (5.6) with $M = -3$ and $N = 2$:

$$X(m) = \begin{cases} 5 & m = 0, \pm 8, \ldots \text{ as before} \\ \frac{e^{3\pi jm/4} - e^{-2\pi jm/4}}{1 - e^{-\pi jm/4}} & \text{otherwise} \end{cases}$$

Hence, except for $m = 0, \pm 8, \ldots$:

$$\begin{aligned} X(m) &= \frac{e^{\pi jm/8}}{e^{\pi jm/8}}\left(\frac{e^{3\pi jm/4} - e^{-2\pi jm/4}}{1 - e^{-\pi jm/4}}\right) \\ &= \frac{e^{7\pi jm/8} - e^{-3\pi jm/8}}{e^{\pi jm/8} - e^{-\pi jm/8}} \\ &= e^{\pi jm/4}\left(\frac{\sin(5\pi m/8)}{\sin(\pi m/8)}\right) \end{aligned}$$

which also correctly yields $X(0) = 5$. Thus, the magnitude spectrum is exactly as before (example 2). However, the effect of the left-shift is seen in the phase spectrum. Whereas this was previously identically zero, it is now:

$$\Phi(m) = m\pi/4$$

which is linear.

5.4 THE INVERSE DFT

The DFT is given by equation (5.4) as:

$$X(m) = \mathcal{D}[x(n)] = \sum_{n=0}^{N-1} x(n)e^{-2\pi jmn/N}$$

which enables us to analyse a real time-domain sequence $x(n)$ into a set of cosine and sine components (or, equivalently, real and imaginary components) spaced by $2\pi/N$ in frequency index, or $2\pi/NT$ in hertz frequency.

Let us multiply both sides of the above equation by $e^{2\pi jlm/N}$ with integer l, and sum from $m = 0$ to $m = (N - 1)$:

$$\sum_{m=0}^{N-1} X(m)e^{2\pi jlm/N} = \sum_{m=0}^{N-1}\sum_{n=0}^{N-1} x(n)e^{-2\pi j(n-l)m/N}$$

Reversing the order of summation on the right-hand side:

$$\sum_{m=0}^{N-1} X(m)e^{2\pi jlm/N} = \sum_{n=0}^{N-1} x(n)\sum_{m=0}^{N-1} e^{-2\pi j(n-l)m/N} \tag{5.7}$$

Using the now-familiar summation formula for a truncated geometric series:

$$\sum_{m=0}^{N-1} e^{-2\pi j(n-l)m/N} = \begin{cases} N & (n-l) = 0, \pm N, \pm 2N, \ldots \\ 0 & \text{otherwise} \end{cases}$$

Hence, the right-hand side of equation (5.7) is only non-zero for $l = n \pm kN$ (or, equivalently, $n = l \pm kN$) where k is any integer, and:

$$\sum_{m=0}^{N-1} X(m)e^{2\pi j(n\pm kN)m/N} = \sum_{n=0}^{N-1} x(n)N \qquad n = l \pm kN$$

Now, there is only one value of n in the range $0 \leq n \leq (N-1)$: namely, $n = l$, $k = 0$. Hence, we can dispense with the summation on the right-hand side. Putting $k = 0$ on the left-hand side then yields the inverse discrete Fourier transform (IDFT):

$$x(n) = \mathcal{D}^{-1}[X(m)] = \frac{1}{N}\sum_{m=0}^{N-1} X(m)e^{2\pi jmn/N} \tag{5.8}$$

The IDFT allows us to synthesise a time-domain sequence $x(n)$ as the addition of a set of complex exponentials (cosinusoids and sinusoids spaced by $2\pi/N$ in frequency) each 'scaled' by $X(m)$ – which is, of course, complex for real $x(n)$.

The DFT and IDFT defining equations (5.4) and (5.8) constitute a transform pair, corresponding to the analysis and synthesis equations respectively. In our development, we have included the $1/N$ multiplier with the IDFT but it is easily possible to formulate an equivalent pair of equations with this multiplier included with the DFT itself. The reader should note that both formulations are seen in the literature: the choice between them is no more than a matter of taste.

Example 5: Find the DT signal whose spectrum in the band $0 \leq m \leq (N-1)$ (c.f. example 2) is:

$$X(m) = \begin{cases} N/2 & m = 1, (N-1) \\ 0 & \text{otherwise} \end{cases} \qquad 0 \leq m \leq (N-1)$$

From the IDFT defining equation (5.8):

$$\begin{aligned} x(n) &= \frac{1}{N}\sum_{m=0}^{N-1} X(m)e^{2\pi jmn/N} \\ &= \frac{1}{N}\left(X(1)e^{2\pi jn/N} + X(N-1)e^{2\pi jn(N-1)/N}\right) \end{aligned}$$

$$= \frac{1}{2}\left(e^{2\pi jn/N} + e^{2\pi jn(N-1)/N}\right)$$
$$= \cos(2\pi n/N)$$

This, of course, is the expected answer in light of example 2. From this example, it should also be clear how the $1/N$ scaling factor can be included with either the forward or the inverse DFT.

Example 6: Find the IDFT of $X(m) = 1$ for all m in the range $0 \leq m \leq (N-1)$.

Substituting into equation (5.8):

$$x(n) = \frac{1}{N}\sum_{m=0}^{N-1} e^{2\pi jmn/N}$$

Applying the summation formula:

$$Nx(n) = \begin{cases} N & n = 0, \pm N, \ldots \\ \frac{1-e^{2\pi jn}}{1-e^{2\pi jn/N}} & \text{otherwise} \end{cases}$$

There are now two ways of proceeding to evaluate $x(n)$ for $n \neq 0, \pm N, \ldots$. First, we could recognise that $e^{2\pi jn/N}$ is an Nth root of unity, i.e. $e^{2\pi jn} = 1$, so that:

$$\begin{aligned} x(n) &= \begin{cases} 1 & n = 0, \pm N, \ldots \\ 0 & \text{otherwise} \end{cases} \\ &= \delta(n - kN) \quad \text{(by definition of the Kronecker delta function)} \end{aligned}$$

This takes the value δ_0 within the range $0 \leq n \leq (N-1)$, and is periodic with period N.

An alternative approach is to ignore (or fail to spot!) the fact that $e^{2\pi jn/N}$ is an Nth root of unity. Then, for $n \neq 0, \pm N, \ldots$:

$$\begin{aligned} Nx(n) &= \frac{e^{\pi jn}\left(e^{-\pi jn} - e^{\pi jn}\right)}{e^{\pi jn/N}\left(e^{-\pi jn/N} - e^{\pi jn/N}\right)} \\ x(n) &= e^{\pi jn(N-1)/N}\frac{1}{N}\left(\frac{\sin(\pi n)}{\sin(\pi n/N)}\right) \end{aligned} \tag{5.9}$$

Given that the spectrum $X(m)$ is rectangular (equal to 1 for all m in the length-N window), and in light of example 3, we might well have expected to be able to express

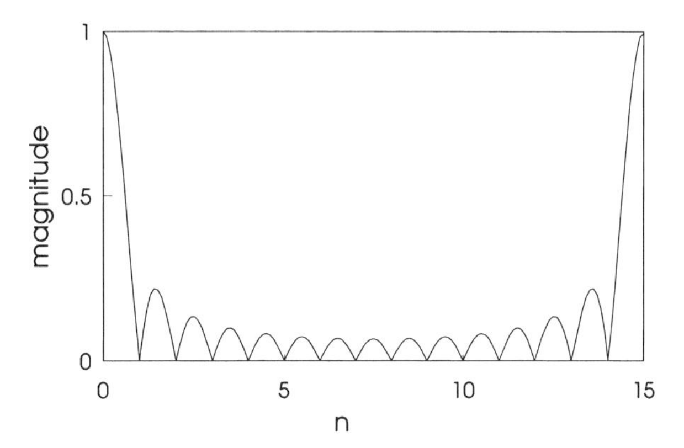

Fig. 5.4 The IDFT of $X(m) = 1$ for all m in a length-N window is a Dirichlet function whose magnitude is shown here for $N = 15$. The function takes zero values when the time index n is integer except for $n = 0, \pm N, \pm 2N, \ldots$.

$x(n)$ in just this form – involving a Dirichlet function. As we shall see very shortly, however, the DFT and its inverse are unique. That is, we cannot have two different inverses for the same $X(m)$. Let us then show that this form, equation (5.9), is precisely equal to $x(n) = \delta_0$ $(0 \leq n \leq (N-1))$.

First, for $n = 0$ and $n = N$, $x(n)$ is purely real. For integer n, the $\sin(\pi n)$ term in equation (5.9) will be zero. So, except when the $\sin(\pi n/N)$ denominator is also zero, $x(n)$ disappears. (The fact that it has non-zero phase is then irrelevant.) When the denominator $\sin(\pi n/N) = 0$, which occurs for $n = 0, \pm N, \ldots$, then $x(n)$ can be evaluated using l'Hôpital's rule (exercise 5.5), to give:

$$x(n) = \delta(n - kN) = \delta_0 \qquad 0 \leq n \leq (N-1)$$

as before.

Figure 5.4 plots the magnitude of equation (5.9), from which it is easily seen that $x(n)$ is zero for integer values of n except $n = 0, \pm N, \pm 2N, \ldots$.

Two points of interest can be made about this result. The first is more or less unexceptional. Examples 1 and 6 together confirm that, although $x(n)$ is an isolated Kronecker delta function, the DFT represents it over all n (in a way that has now become familiar) as a periodic function with period N. The second point is somewhat more remarkable. Although $x(n) = \delta_0$ is not band-limited, its inverse is obtained from its DFT spectrum without aliasing. This is because, as we have seen, the Fourier transform represents the 'underlying' CT signal $x(t) = \delta(t)|_{t=0}$ as a Dirichlet function. It happens that – with the exception of the very points $n = \pm kN$ where the delta function is non-zero – the DFT samples the underlying function at precisely those points where the Dirichlet function is zero.

5.5 FURTHER PROPERTIES OF THE DFT

We have studied the periodicity properties of the DFT in some depth. The DFT has many other important properties, however, several of which we have already encountered. In this section, we review these properties and relate them to our earlier studies.

5.5.1 Uniqueness property

In the IDFT for $X(m)$ (equation (5.8)), the exponential term $e^{2\pi jmn/N}$ is periodic in m with period N. Consequently, $X(m)$ is single-valued within the range of the window $0 \leq m \leq (N-1)$. Within this range, $x(n)$ is recovered uniquely using the inverse DFT as a synthesis equation. The periodicity of the exponential term means that outside this range, the sequence $x(n)$ repeats with period N. If we interpret the index n as modulo-N, the uniqueness property is maintained.

Since the forward DFT, equation (5.4), is similar in form – involving an exponential term $e^{-2\pi jmn/N}$ – the uniqueness property holds for this too. That is, given a particular $x(n)$, the DFT is found uniquely.

5.5.2 Time-frequency duality

There is a remarkable similarity in the definitions of the forward and inverse DFT, equations (5.4) and (5.8). They differ only in the sign of the exponent and in the scaling factor $1/N$. Otherwise, they are identical. This fact has important consequences. First, from a theoretical point of view, it signifies that there is a duality between the time and frequency domains. Hence, if a DT signal (e.g. a sine wave) is extensive in time and has a compact spectrum, then it follows that a compact spectrum will correspond to a signal which is extensive in time. From a practical point of view, because the mechanics of computing the DFT and IDFT are so similar, it is possible to use essentially the same algorithm for doing both.

One of the important aspects of time-frequency duality has been central to our discussion of the DFT. We are familiar with the fact that sampling of a time-domain signal leads to a frequency-domain representation which is periodic in frequency. By duality, sampling of the Fourier spectrum of a signal (as the DFT does) leads to a time-domain representation which is inherently periodic.

5.5.3 Linearity

The linearity property is stated as:

$$\begin{aligned}
\text{if} \quad X(m) &= \mathcal{D}[x(n)] \\
\text{and} \quad Y(m) &= \mathcal{D}[y(n)] \\
\text{then } \mathcal{D}[ax(n) + by(n)] &= a\mathcal{D}[x(n)] + b\mathcal{D}[y(n)] \\
&= aX(m) + bY(m)
\end{aligned}$$

where $x(n)$ and $y(n)$ are equal length sequences.

Again, this is a statement of the superposition principle. This property follows directly from the definition of the DFT. Because of the essential similarity of the DFT and IDFT defining equations, an obvious dual property holds:

$$\begin{aligned}
\text{if} \quad x(n) &= \mathcal{D}^{-1}[X(m)] \\
\text{and} \quad y(n) &= \mathcal{D}^{-1}[Y(m)] \\
\text{then } \mathcal{D}^{-1}[aX(m) + bY(m)] &= a\mathcal{D}^{-1}[X(m)] + b\mathcal{D}^{-1}[Y(m)] \\
&= ax(n) + by(n)
\end{aligned}$$

5.5.4 Shift and modulation properties

Consider the sequence $y(n) = x(n-k)$ produced from $x(n)$ by a right-shift of k samples. By definition of the DFT:

$$\begin{aligned}
\mathcal{D}[y(n)] &= \sum_{n=0}^{N-1} y(n)e^{-2\pi jmn/N} \\
\text{and} \quad \mathcal{D}[x(n-k)] &= \sum_{n=0}^{N-1} x(n-k)e^{-2\pi jmn/N} \qquad (5.10)
\end{aligned}$$

Put $(n-k) = \Delta$, so that:

$$\mathcal{D}[x(n-k)] = \sum_{\Delta=-k}^{N-1-k} x(\Delta)e^{-2\pi j(k+\Delta)m/N}$$

$$= e^{-2\pi jmk/N} \sum_{\Delta=-k}^{N-1-k} x(\Delta) e^{-2\pi j \Delta m/N}$$

Here, the limits of the summation are outside of the $0 \le \Delta \le (N-1)$ window. However, we are now familiar with the periodicity properties of the DFT, whereby the representation implicitly assumes the sequence $x(\Delta)$ to be periodic for all Δ with period N. Hence:

$$\sum_{\Delta=-k}^{N-1-k} x(\Delta) e^{-2\pi j \Delta m/N} = \sum_{\Delta=0}^{N-1} x(\Delta) e^{-2\pi j \Delta m/N}$$

$$\text{and} \quad \mathcal{D}[x(n-k)] = e^{-2\pi jmk/N} X(m)$$

This is the right-shift property of the DFT. It was illustrated in example 4 above where we left-shifted the rectangular pulse by one sample period (i.e. a right-shift of $k = -1$).

The dual property is of some interest. Multiplying the time-domain signal by a complex exponential will yield a shifted spectrum. That is:

$$\mathcal{D}\left[x(n) e^{2\pi jnk/N}\right] = X(m-k)$$

This is the *modulation property*. Multiplication by a 'carrier' of frequency $2\pi k/T$ shifts the spectrum of the modulated signal $x(n)$ to the location of the carrier.

Do not be disturbed by the fact that the time-domain sequence above is complex. This is just a way of denoting that multiplication by a carrier sinusoid of any phase will suffice to modulate the signal.

5.5.5 Convolution properties

The now familiar convolution-sum of two sequences is:

$$x(n) * y(n) = \sum_{k=-\infty}^{\infty} x(k) y(n-k)$$

Now, if $x(n)$ and $y(n)$ are each length-N, their convolution is of length $(2N-1)$. But in using the DFT, we are restricted to length-N sequences. This leads to the idea of *circular convolution*, whereby we consider the ends of the window for each sequence to be 'joined'. That is, the relative shift between the two is evaluated modulo-N. The result of circular convolution is then a length-N sequence.

Consider the (circular) convolution of two sequences:

$$x(n) * y(n) = \sum_{k=0}^{N-1} x(k)y(n-k)$$

So:

$$\begin{aligned} \mathcal{D}[x(n) * y(n)] &= \sum_{n=0}^{N-1}\sum_{k=0}^{N-1} x(k)y(n-k)e^{-2\pi jmn/N} \\ &= \sum_{k=0}^{N-1} x(k) \sum_{n=0}^{N-1} y(n-k)e^{-2\pi jmn/N} \end{aligned}$$

But from the shift property above:

$$\begin{aligned} \mathcal{D}[y(n-k)] &= Y(m)e^{-2\pi jkm/N} \\ &= \sum_{n=0}^{N-1} y(n-k)e^{-2\pi jmn/N} \qquad \text{(see equation (5.10) above)} \end{aligned}$$

Hence:

$$\begin{aligned} \mathcal{D}[x(n) * y(n)] &= \sum_{k=0}^{N-1} x(k)Y(m)e^{-2\pi jkm/N} \\ &= X(m)Y(m) \end{aligned}$$

This is the very important *time-convolution property*, whereby convolution in the time domain can be effected by multiplication in the frequency domain.

Let us consider the dual property. That is, we wish to know the effect in the time domain of convolving spectra $X(m)$ and $Y(m)$ in the frequency domain. Now:

$$X(m) * Y(m) = \sum_{k=0}^{N-1} X(k)Y(m-k)$$

By definition of the IDFT:

$$\begin{aligned}
\mathcal{D}^{-1}[X(m) * Y(m)] &= \frac{1}{N}\sum_{n=0}^{N-1}\sum_{k=0}^{N-1} X(k)Y(m-k)e^{2\pi jmn/N} \\
&= \frac{1}{N}\sum_{k=0}^{N-1} X(k)\sum_{n=0}^{N-1} Y(m-k)e^{2\pi jmn/N}
\end{aligned}$$

But by the modulation property:

$$\begin{aligned}
\mathcal{D}^{-1}[Y(m-k)] &= y(n)e^{2\pi jnk/N} \\
&= \frac{1}{N}\sum_{n=0}^{N-1} Y(m-k)e^{2\pi jmn/N}
\end{aligned}$$

Hence:

$$\begin{aligned}
\mathcal{D}^{-1}[X(m) * Y(m)] &= \sum_{k=0}^{N-1} X(k)y(n)e^{2\pi jnk/N} \\
&= N[x(n)y(n)]
\end{aligned}$$

This is the *frequency-convolution property* – convolution in the time domain corresponds to multiplication in the frequency domain and, by duality, *vice versa*. Obviously, the factor N which appears in the dual convolution property must not be forgotten!

5.5.6 Symmetry properties

The DFT of a real sequence has, in common with other Fourier representations of DT signals, some significant symmetry properties. These are important for a proper understanding of the DFT. Additionally, the symmetry properties have practical importance: they can be exploited in algorithmic implementations to increase efficiency.

The real sequence $x(n)$ with $0 \leq n \leq (2M-1)$, i.e $M = N/2$ has DFT:

$$X(m) = \sum_{n=0}^{2M-1} x(n)e^{-2\pi jmn/2M}$$

Hence, for $0 \leq k \leq (M-1)$:

$$X(M+k) = \sum_{n=0}^{2M-1} x(n)e^{-2\pi jn(M+k)/2M}$$

Since $e^{2\pi jn} = 1$ for all integer n, we can multiply the summand by $e^{4\pi jnM/2M}$ to give:

$$\begin{aligned} X(M+k) &= \sum_{n=0}^{2M-1} x(n)e^{4\pi jnM/2M}e^{-2\pi jn(M+k)/2M} \\ &= \sum_{n=0}^{2M-1} x(n)e^{2\pi jn(M-k)/2M} \end{aligned}$$

But the right-hand side of this equation is the complex conjugate of A, so:

$$X(M+k) = \tilde{X}(M-k)$$

This is the *complex-conjugate property* of the DFT, which we previously met when studying the frequency response of a DT system. It has several important consequences which we outline below. Before doing so, however, let us look more closely at the first DFT coefficient $X(0)$:

$$X(0) = \sum_{n=0}^{2M-1} x(n)$$

which, for real $x(n)$, is real. In the formulations of the DFT which include the $1/N$ multiplier with the forward transform, it can be directly interpreted as the average ('dc') value of the sequence. In the formulation to which we are working, $X(0)$ is the average value scaled by N.

Consider next the 'centre' value $X(M)$:

$$\begin{aligned} X(M) &= \sum_{n=0}^{2M-1} x(n)e^{-2\pi jnM/2M} \\ &= \sum_{n=0}^{2M-1} x(n)(-1)^n \end{aligned} \tag{5.11}$$

which is also real. Because $m = M = N/2$ is the Nyquist frequency, we might expect that $X(M)$ will be zero for a sequence representing a periodic, band-limited CT signal

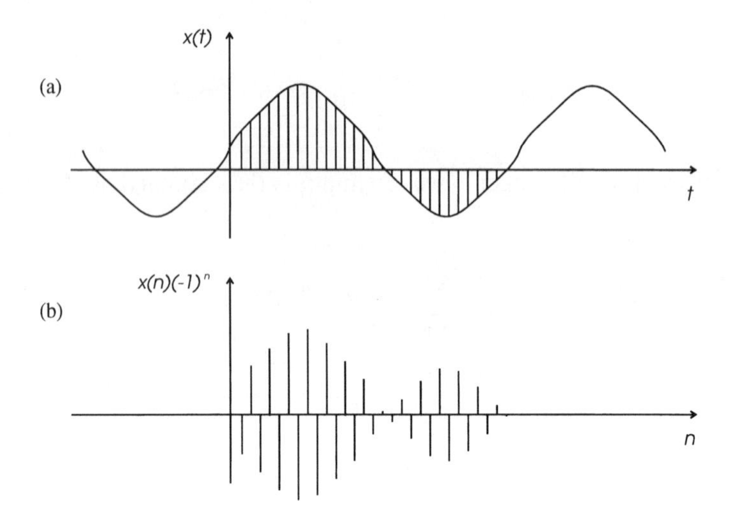

Fig. 5.5 For a periodic CT signal as in (a) properly sampled with an integer number of periods in the analysis window (just one here), the sequence $x(n)(-1)^n$ as in (b) will sum to zero.

sampled at a rate above the (minimum) Nyquist rate. That is, there will be no non-zero DFT coefficient for $n = M$ because the signal is band-limited to a frequency below:

$$M \times \frac{2\pi}{NT} = \frac{\pi}{T} \qquad \text{with } M = N/2$$

Now, equation (5.11) states that $X(M)$ is found by summing a sequence produced from $x(n)$ by changing the sign of alternate sample values as depicted in figure 5.5 for a single period. Intuitively, it seems reasonable that, provided there are an integer number of periods in the analysis window, the sum will be zero. If sampled at a sufficiently high rate, adjacent samples will differ very little. Thus, the positive sample values will represent a half-cycle of the period just as well as the negative samples with which they alternate, and their sum over the half-cycle will be zero.

From the above, it is apparent that we only need the first M $(= N/2)$ coefficients to specify the DFT of a real sequence fully. One of these, $X(0)$, is real and is a measure of the average ('dc') value of the sequence. The other $(M - 1)$ values are complex. If $x(t)$ is periodic, band-limited and properly sampled, and there are an integer number of periods in the analysis window, the 'centre' value, $X(M)$, will be identically zero. However, we have specifically developed the DFT to be applicable to aperiodic signals. Hence, we will generally need to include $X(M)$. The remaining $(M - 1)$ (complex) coefficients are the complex conjugates of the values $X(1)$, $X(2)$, ..., $X(M - 1)$. In total, then, we need 2 real and $(M-1) = (N/2-1)$ complex coefficients, corresponding to $2+2(N/2-1) = N$ independent values, to represent a real length-N sequence. That is, the same number of values are required to specify the DT signal irrespective of whether we represent it in the time or the frequency domain.

Because of the complex-conjugate property, symmetry relations hold for the DFT such that:

$$|X(M+k)| = |X(M-k)|$$

and:

$$\angle X(M+k) = -\angle X(M-k)$$

that is, the magnitude and phase responses have even and odd symmetry respectively about the Nyquist point. This is a more general statement of the symmetry properties we found in our earlier study of the frequency response of a DT system (chapter 3). Those properties held precisely because the frequency response of a DT system is the Fourier transform of the system's impulse response.

It is, of course, possible to write the DFT as:

$$X(m) = \sum_{n=0}^{N-1} x(n)\,[\cos(2\pi mn/N) - j\sin(2\pi mn/N)]$$

Now, cos( ) is an even function of frequency while sin( ) is an odd function. Hence, if $x(n)$ is real, the symmetry relations we have discovered are only to be expected: but if $x(n)$ is complex, they do not hold. We need $2N$ degrees of freedom in specifying the input and, correspondingly, $2N$ degrees of freedom in specifying the spectrum. Thus, all N complex spectral coefficients are required.

5.6 PARSEVAL'S THEOREM AND DT SIGNALS

In section 1.7, we defined the average power in a (periodic) finite-power DT signal as:

$$P_{av} = \lim_{N\to\infty} \frac{1}{2N+1} \sum_{n=-N}^{N} x^2(n)$$

For a right-sided signal, we can write:

$$P_{av} = \lim_{N\to\infty} \frac{1}{N} \sum_{n=0}^{N-1} x^2(n)$$

If we are able to assume that an integer number of periods of $x(t)$ are represented in the window $0 \leq n \leq (N-1)$, then:

$$P_{av} = \frac{1}{N}\sum_{n=0}^{N-1} x^2(n)$$

which is the average power in the time domain. In principle, we can always make this assumption, since the DFT treats the signal in the analysis window as just one period from a periodic $x(t)$ that extends over all time (see below). But how do we compute the power from the spectral representation? We first note that:

$$\begin{aligned} \mathcal{D}[x^2(n)] &= \sum_{n=0}^{N-1} x^2(n) e^{-2\pi jmn/N} \\ \text{so that} \quad P_{av} &= \frac{1}{N}\mathcal{D}[x^2(n)]\Big|_{m=0} \end{aligned}$$

Now, from the frequency-convolution property:

$$\begin{aligned} \mathcal{D}[x(n)x(n)] &= \frac{1}{N} X(m) * X(m) = \frac{1}{N}\sum_{k=0}^{N-1} X(k)X(m-k) \\ \text{and} \quad P_{av} &= \frac{1}{N^2}\sum_{k=0}^{N-1} X(k)X(m-k)\Bigg|_{m=0} \\ &= \frac{1}{N^2}\sum_{k=0}^{N-1} X(k)X(-k) \end{aligned}$$

From the complex-conjugate property:

$$\begin{aligned} X(-k) &= \tilde{X}(k) \\ \text{and} \quad X(k)X(-k) &= |X(k)|^2 \end{aligned}$$

so that:

$$P_{av} = \frac{1}{N}\sum_{n=0}^{N-1} x^2(n) = \frac{1}{N^2}\sum_{k=0}^{N-1} |X(k)|^2$$

This is another specific version of Parseval's theorem which we met in section 1.7. It states that the average power can be evaluated either in the time or the frequency domain, by virtue of the equality:

$$\sum_{n=0}^{N-1} x^2(n) = \frac{1}{N}\sum_{m=0}^{N-1} |X(m)|^2 \tag{5.12}$$

The representation $|X(m)|^2$ is called the *power spectrum* of $x(n)$. Note that the power spectrum depends only upon the magnitude spectrum and is independent of the phase spectrum.

Example 7: Confirm Parseval's theorem for the case of the rectangular pulse in example 3.

The pulse is specified as:

$$x(n) = \{1, 1, 1, 0, 0, 0, 1, 1\}$$

So:

$$\sum_{n=0}^{7} p_n^2 = 5$$

The DFT spectrum is given as the Dirichlet function:

$$X(m) = \frac{\sin(5\pi m/8)}{\sin(\pi/8)}$$

yielding the sequence (real in this case) for $0 \leq m \leq 7$:

$$X(m) = \{5, 2.414214, -1, -0.41421, 1, -0.41421, -1, 2.414214\} \tag{5.13}$$

The reader should note in passing the symmetry of this sequence about the point $m = M = N/2 = 4$ which confirms our expectations from section 5.5.6. Note too that $X(4)$ is non-zero ($= 1$) reflecting the fact that the rectangular pulse is not a band-limited signal. From equation (5.13):

$$\begin{aligned}\frac{1}{N}\sum_{n=0}^{7} |X(m)|^2 &= \frac{1}{8}(25 + 5.828427 + 0.171573 + 1 \\ &\qquad + 0.171573 + 5.828427 + 25) \\ &= 5\end{aligned}$$

as required.

On page 148, we explored two possible ways of interpreting the DFT. First, if $x(n)$ is truly periodic, then the $X(m)$ values are proportional to the coefficients of the discrete Fourier series. In this case, $x(n)$ will generally be finite-power, it will have a line spectrum, and the power spectrum $|X(m)|^2$ in equation (5.12) is an entirely appropriate representation. Our second interpretation, however, considered the more likely case that $x(n)$ is aperiodic. Then, the discrete-in-frequency nature of the DFT means that we are actually treating the length-N window of $x(n)$ as if it were one period of a periodic signal. According to this interpretation, we can view $x(n)$ as if it were finite-power and the $X(m)$ values as defining its power spectrum, $|X(m)|^2$. Alternatively, we can maintain our view of $x(n)$ as aperiodic and finite-energy by treating the $X(m)$ values as samples of the underlying continuous spectrum $X(j\omega)$. That is, the $|X(m)|^2$ values of the power spectrum are samples of the underlying energy density spectrum $|X(j\omega)|^2$.

5.7 Truncation and windowing

The DFT represents the spectrum of a CT signal $x(t)$ by analysing a finite number of samples, N, of the signal. In effect, the DFT samples the 'underlying' spectrum $X(j\omega)$ of $x(t)$ at the minimum (Nyquist) rate, with a frequency spacing of $2\pi/NT$. Thus far, we have assumed that the signal is *deterministic*, i.e. it can be described by a relatively simple formula, and that N is reasonably small. More usually, however, $x(t)$ is non-deterministic and extends widely over time so that it is just not practical to process all of it. It has to be broken down into a series of truncated (length-N) *blocks* for analysis. In general then, $x(t)$ will extend outside of the length-N (duration NT) window and the obtained DFT spectrum will be only an approximation to the desired spectrum. The evaluation of the DFT when $x(t)$, and thereby $x(n)$, is non-deterministic is the subject of chapter 7.

Consider figure 5.6(a) which shows a single block of a sampled sine wave truncated to the range $0 \leq n < 50$. The truncation process corresponds to multiplication by a rectangular window, $w(n)$:

$$w(n) = \begin{cases} 1 & 0 \leq n < 50 \\ 0 & \text{otherwise} \end{cases}$$

In (b), we see the periodic signal formed by repeating the windowed sinusoid with period $N = 50$. It is apparent that the waveform in figure 5.6 is rather different from a pure sinusoid, yet it is this waveform whose spectrum we are actually finding with the DFT. Clearly, as a result of the step discontinuities, this will have a more wideband spectrum that the pure sinusoid. The resulting spreading of spectral energy is called

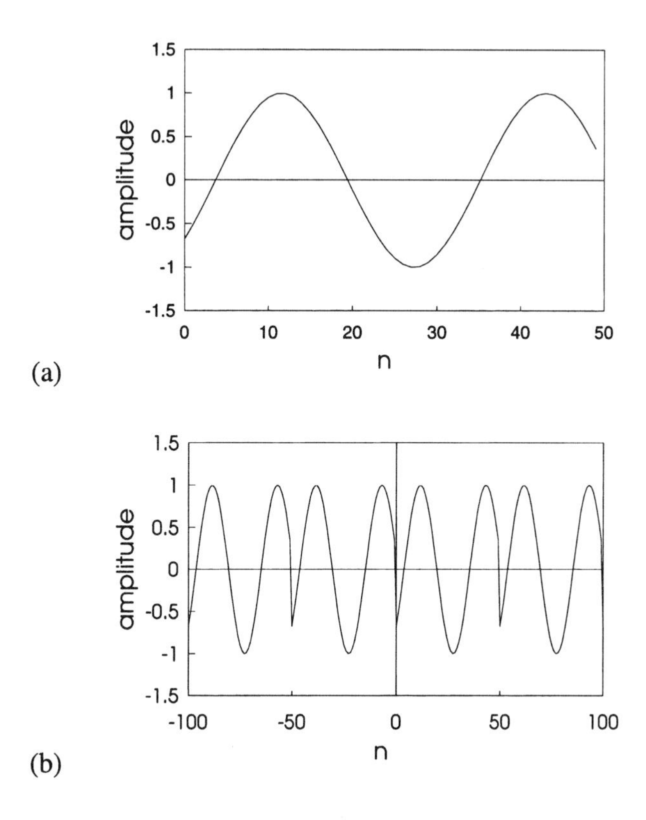

Fig. 5.6 (a) Single block of a sine wave truncated by multiplication by a length-50 rectangular window. (b) Periodic function having single period as in (a).

spectral leakage. Is there any way that we can reduce such leakage, so improving our estimate of the spectrum of the underlying sequence?

One possibility might be to fit a linear trend to the data sequence, and then to remove it by subtraction before analysis. This can work well in some circumstances – especially those where the data have indeed been contaminated by such a linear trend. This commonly happens as a result of drift in the dc gain of a signal-conditioning amplifier. Another possibility is to reduce the discontinuities at the ends of the analysis window by using a window function other than rectangular. Many popular window functions smooth the data to a value close to zero at $n = 0$ and $(N - 1)$ rather than truncating it abruptly. One much-used function is the Hamming (raised cosine) window:

$$w(n) = 0.54 - 0.46\cos\left(\frac{2\pi n}{N-1}\right) \tag{5.14}$$

Figure 5.7 (a) shows a length-50 Hamming window function. In figure 5.7 (b), we show the effect of multiplying the truncated sine wave of figure 5.6 (a) by this window. Figure 5.7 (c) depicts the periodic signal having the windowed function in (b) as its basic period. Although this is still some way from being a pure sinusoid, at least the severe step discontinuities of figure 5.6 (b) have been considerably reduced so, in turn, reducing spectral leakage.

The 'full' Hamming window of figure 5.7 (a) only reaches the value 1 at the centre point and, hence, has a rather severe effect overall. That is, data well away from the ends of the analysis window are significantly attenuated. A frequently-used palliative is to apply the raised cosine function only at the ends of the window – to a fixed percentage of the data. Figure 5.8 shows the effect of using the function to taper only the 10% of the data (here 5 samples) at each end of the window. Intuitively, it appears that we have been reasonably successful in reducing the step discontinuities (and, hence, spectral leakage) at the window boundaries, without distorting the waveform we are attempting to analyse too dramatically.

Our study of the properties of the DFT tells us that multiplication of a time domain signal by a windowing function corresponds to convolution of the signal's spectrum with that of the window. Hence, determining the exact effect of using a particular window function requires us to examine its spectrum. From example 6, where we considered the IDFT of a rectangular spectrum, we expect (by the principle of time-frequency duality) the spectrum of a rectangular window to be a Dirichlet function also. To confirm this:

$$\begin{aligned}
W(m) &= \sum_{n=0}^{N-1} e^{-2\pi jmn/N} \\
&= \frac{1-e^{-2\pi jm}}{1-e^{-2\pi jm/N}} \\
&= \frac{e^{-\pi jm}\left(e^{\pi jm}-e^{-\pi jm}\right)}{e^{-\pi jm/N}\left(e^{\pi jm/N}-e^{-\pi jm/N}\right)} \\
&= e^{-\pi jm(N-1)/N}\left(\frac{\sin(\pi m)}{\sin(\pi m/N)}\right) \qquad (5.15)
\end{aligned}$$

The similarity to equation (5.9) is clear. Thus, the magnitude spectrum of the rectangular window will be very much as shown in figure 5.4 (except for the $1/N$ scaling

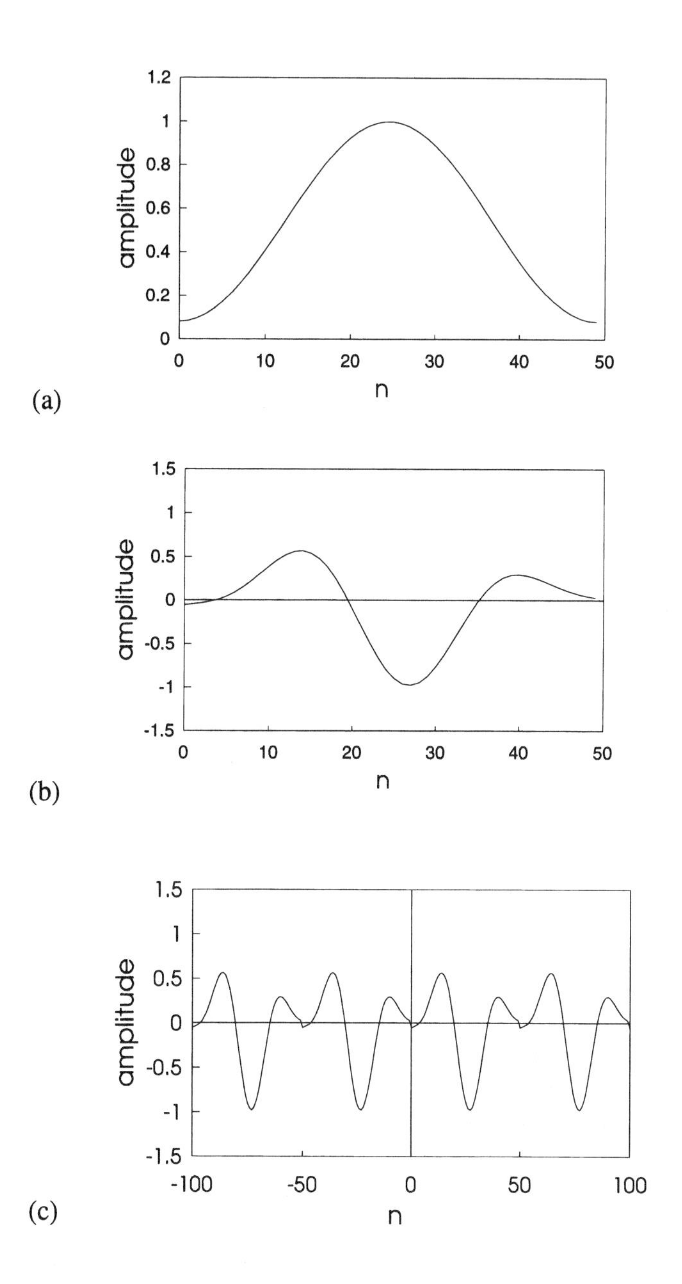

Fig. 5.7 (a) Length-50 Hamming window. (b) Sine wave of figure 5.6 (a) multiplied by this window. (c) Periodic function having single period as in (b).

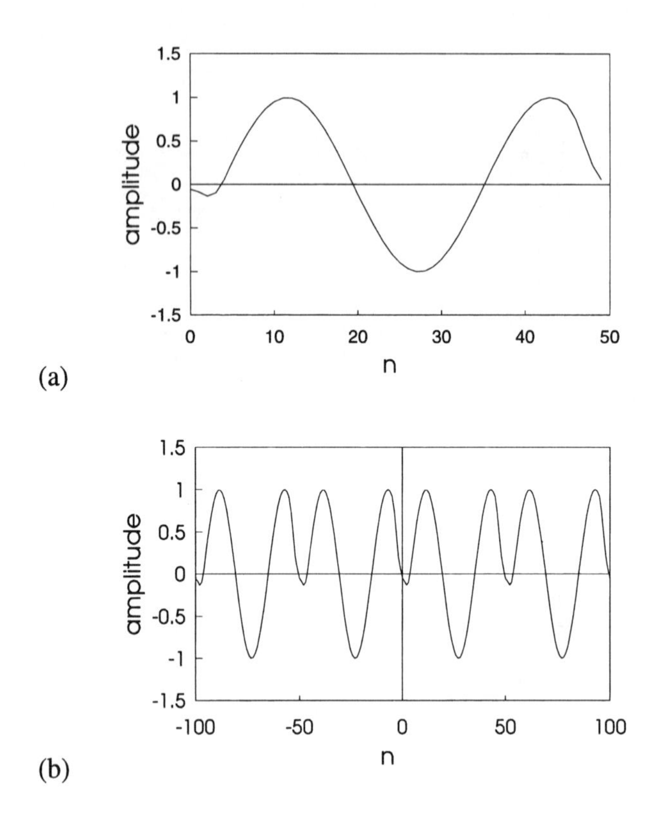

Fig. 5.8 (a) Sine wave truncated by multiplication by a 10% Hamming taper. (b) Periodic function having single period as in (a).

factor and the change of discrete variable from n (time) to m (frequency)).

Again, the zero values of this Dirichlet function occur exactly at the integer values of m, which means that it aliases a Kronecker delta function (but this time of height N). Now, convolution with a Kronecker delta function is an identity operation. In other words, when performing the convolution, all the sample points of the signal spectrum – except just one corresponding to the peak – align with the zero points of the window spectrum, whether we consider this to be a Dirichlet function or (equivalently) a delta function. Hence, although multiplication by the rectangular window truncates the signal whose spectrum we are estimating, the only effect on the spectrum of the *truncated* signal is to scale it by N. In this sense, the rectangular window is an implicit part of the DFT.

Additional insight can be gained into the spectral properties of the rectangular win-

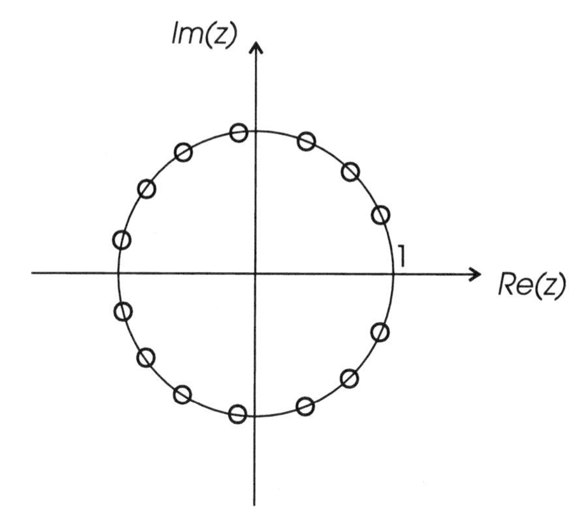

Fig. 5.9 The zeros of the z-transform of the rectangular window $w(n) = 1$ for all n in the range $0 \leq n \leq (N-1)$ are equally spaced around the unit circle. There is cancellation of the zero at $z = 1$ with the single pole at the same point. This is shown for the case of $N = 15$.

dow by considering its z-transform:

$$
\begin{aligned}
W(z) &= \sum_{n=0}^{N-1} z^{-n} \\
&= \frac{1 - z^{-N}}{1 - z^{-1}}
\end{aligned}
$$

So, $W(z)$ has zeros which are the Nth roots of unity. In other words, its zeros are equally spaced around the unit circle (figure 5.9). There is, however, a pole at $z = 1$ which cancels the zero at this point, as well as yielding a -6 dB/octave fall-off with frequency. Hence, the spectrum of $w(n)$, which corresponds to the evaluation of the z-transform around the unit circle, has the form shown in figure 5.4 earlier. It has zero value at integer values of m, but is non-zero (equal to 1) at $m = 0$ and $m = N$ corresponding to $z = 0$ and $z = e^{2\pi j}$.

Finding the spectrum of the (full) Hamming window is left as an exercise for the reader (exercise 5.12).

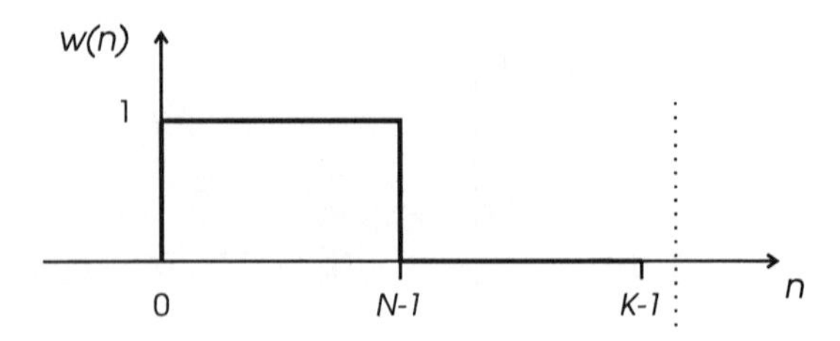

Fig. 5.10 The length-K window implicitly used to zero-pad a length-N sequence is rectangular (equal to 1) in the range $0 \le n \le (N-1)$ and zero in the range $N \le n \le (K-1)$.

5.8 INTERPOLATION BY ZERO-PADDING

One of the important practical applications of the DFT is interpolation to increase the resolution of the signal representation, either in the frequency or the time domain. This is done by the operation of zero-padding, which takes a sequence (or block) of length N and produces a sequence of length K ($> N$) by adding $(K - M)$ zeros.

Let us consider the effect of zero-padding the time-domain representation $x(n)$ on the frequency-domain representation, $X(m)$. If the underlying CT signal $x(t)$ is properly sampled, all relevant information about it is present and is preserved by the zero-padding. The same is true of the spectral representation, since the DFT properly samples the underlying spectrum of $x(t)$. The increase in sequence length, however, reduces the sample spacing in the frequency domain from $2\pi/NT$ to $2\pi/KT$, thereby increasing the resolution. By this means, we might (for example) improve our estimate of the location of a spectral peak.

Now, zero-padding corresponds implicitly to multiplication by a window:

$$w(n) = \begin{cases} 1 & 0 \le n \le (N-1) \\ 0 & N \le n \le (K-1) \end{cases}$$

Figure 5.10 illustrates this for the case $K = 2N$. The effect of multiplying by $w(n)$ is to convolve the spectrum of the signal being analysed by the window spectrum. When $K = 2N$, this is:

$$W(m) = \sum_{n=0}^{K-1} w(n) e^{-2\pi jmn/K}$$

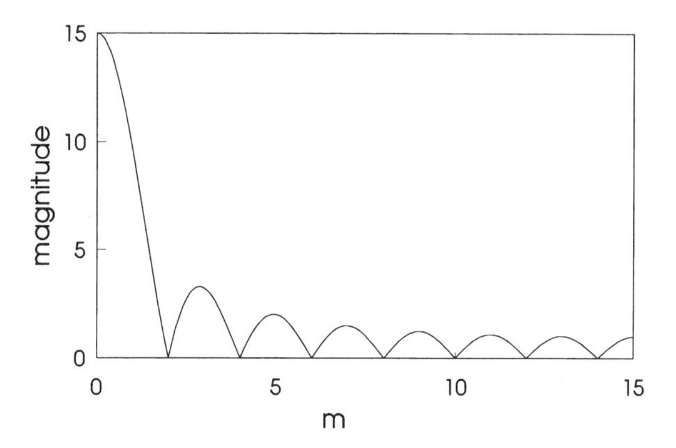

Fig. 5.11 Magnitude spectrum of the zero-padding window of figure 5.10. The spectrum takes zero values when the frequency index m is an even integer, except for $m = 0, \pm N, \pm 2N, \ldots$. It is shown here over a half-period, for the specific case of $K = N/2 = 15$.

$$
\begin{aligned}
&= \sum_{n=0}^{N-1} e^{-\pi j m n/N} \\
&= \frac{1 - e^{-\pi j m}}{1 - e^{-\pi j m/N}} \\
&= \frac{e^{-\pi j m/2}\left(e^{\pi j m/2} - e^{-\pi j m/2}\right)}{e^{-\pi j m/2N}\left(e^{\pi j m/2N} - e^{-\pi j m/2N}\right)} \\
&= e^{-\pi j m(N-1)/2N}\left(\frac{\sin(\pi m/2)}{\sin(\pi m/2N)}\right)
\end{aligned}
$$

Figure 5.11 shows a half-period of the magnitude spectrum of this zero-padding window, for the specific values $K = 30$ and $N = 15$. It has the now-familiar Dirichlet form as a result of the rectangular pulse shape in the time domain. Comparing this spectrum with the IDFT of the rectangular frequency-domain window of figure 5.4, we see that the former takes zero values when the frequency variable m is an even integer (except for $m = 0, \pm N, \ldots$) rather than for all integer values of the time variable n as does the latter. So multiplying in the frequency domain by the length-N rectangular window corresponds to convolution in the time-domain with the Dirichlet function of figure 5.4. In taking this convolution, only the (height-N) peak produces a non-zero product in the convolution-sum. This amounts to the identity operation of convolving with a Kronecker delta function but scaled by N – hence, the $1/N$ factor in the IDFT. By contrast, with zero-padding interpolation, half of the product values in the convolution-sum are non-zero, rather than just one. So instead of being merely an identity-plus-scaling operation, the convolution interpolates values intermediate between those of the original (length-N) sequence.

By the principle of time-frequency duality, we can just as well improve the signal resolution in the time domain by taking the DFT, zero-padding in the frequency domain and taking the inverse DFT.

5.9 Ideal interpolation

Having studied the DFT and, especially, the transform properties of rectangular waves and windows, we are now in a position to answer a question posed in chapter 1: given a properly-sampled signal, how do we reconstitute the underlying CT signal? That is, what is the *ideal* interpolation function to employ in digital-to-analogue conversion?

In section 5.1 above, we found that we could recover $x(t)$ from $x^*(t)$ by passing the latter through an ideal low-pass filter with cut-off B (or, equivalently, ω_s) so as to leave the baseband contribution only (see figure 5.1). Now, the ideal low-pass filter is a rectangular window multiplying $X^*(j\omega)$ in the frequency domain. Hence, we can infer that ideal interpolation of $x^*(t)$ to recover $x(t)$ will correspond to time-domain convolution of $x^*(t)$ with a Dirichlet or sinc function.

For simplicity, we will consider infinite sequences initially. Let $\hat{x}(t)$ be the CT function found from $x^*(t)$ by interpolation. Then:

$$\hat{x}(t) = \mathcal{F}^{-1}[X^*(j\omega)W(j\omega)]$$

$$\text{where} \quad W(j\omega) = \begin{cases} 1 & -\omega_s/2 \leq \omega \leq \omega_s/2 \\ 0 & \text{otherwise} \end{cases}$$

Hence:

$$\begin{aligned} \hat{x}(t) &= \frac{1}{2\pi}\int_{-\infty}^{\infty} X^*(j\omega)W(j\omega)e^{j\omega t}\,d\omega \\ &= \frac{1}{2\pi}\int_{-\omega_s/2}^{\omega_s/2} X^*(j\omega)e^{j\omega t}\,d\omega \end{aligned}$$

From equation (5.1):

$$X^*(j\omega) = \sum_{n=0}^{\infty} x(n)e^{-j\omega nT}$$

so that:

$$\begin{aligned}
\hat{x}(t) &= \frac{1}{2\pi}\int_{-\omega_s/2}^{\omega_s/2}\sum_{n=0}^{\infty} x(n)e^{-j\omega nT}e^{j\omega t}\,d\omega \\
&= \int_{-f_s/2}^{f_s/2}\sum_{n=0}^{\infty} x(n)e^{2\pi jf(t-nT)}\,df \\
&= \sum_{n=0}^{\infty} x(n)\left[\frac{e^{2\pi jf(t-nT)}}{2\pi j(t-nT)}\right]_{-f_s/2}^{f_s/2} \\
&= \sum_{n=0}^{\infty} x(n)\frac{\sin(\pi f_s(t-nT))}{\pi f_s(t-nT)} \\
&= \sum_{n=0}^{\infty} x(n)\frac{\sin(\pi(t-nT)/T)}{\pi(t-nT)/T} \qquad (5.16)
\end{aligned}$$

Hence, as expected, the ideal interpolation is a convolution of $x(n) = x^*(t)$ with a sinc function. Since the interpolation is ideal, we can replace $\hat{x}(t)$ on the left-hand side of equation (5.16) with $x(t)$. At sample instants $t = kT$ in the time domain, the sinc function is zero except for $k = n$.

Figure 5.12 illustrates the interpolation process for the example sequence:

$$\begin{aligned}
x(n) &= \sin[(n+1)\pi/8] \qquad 0 \le n \le 6 \\
&= \{0.3827, 0.7071, 0.9238, 1.0, 0.9238, 0.7071, 0.3827\}
\end{aligned}$$

Convolution with the sinc function corresponds to time reversal of the function (which has no effect because of its symmetry), shifting for all n for which there is overlap, multiplying and adding contributions at all values of t. (In figure 5.12, $x(t)$ (shown dotted) has actually been calculated from the sin( ) function rather than interpolated from equation (5.16), only because the interpolation from just 6 values is rather poor.)

The interpolation equation (5.16) has been developed using $X^*(j\omega)$. This, however, is a continuous-frequency representation. As a result, the convolution is linear

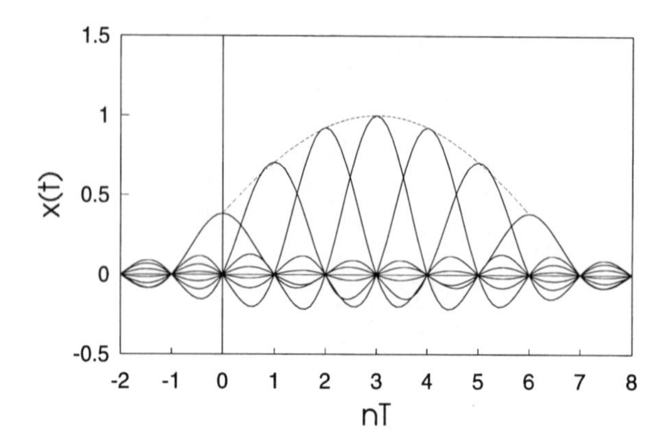

Fig. 5.12 Interpolation of $x(n) = \sin((n+1)\pi/8)$ for $0 \leq n \leq 6$ to give $x(t)$ (shown dotted) using the ideal sinc interpolation function of equation (5.16).

and involves infinite sequences. It should be readily apparent that we can develop a parallel equation based on the DFT and finite-length sequences. This parallel formulation will, of course, involve circular (rather than linear) convolution.

5.10 LINEAR FILTERING USING THE DFT

In linear filtering, we wish to process an input sequence $x(n)$ by passing it through some system with impulse response $h(n)$. The output of this process, $y(n)$, can be found by linear convolution of $x(n)$ with $h(n)$. The DFT gives us a way of replacing the awkward operation of convolution in the time-domain by the simpler operation of multiplication of $X(m) = \mathcal{D}[x(n)]$ by $H(m) = \mathcal{D}[h(n)]$ in the frequency domain, followed by inverse transformation. However, this multiplication is equivalent to circular (rather than linear) convolution of the two time-domain sequences.

Suppose $x(n)$ and $h(n)$ are right-sided sequences of finite length J and K respectively. Now, the linear convolution $x(n) * y(n)$ has length $N = (J + K - 1)$. If we zero-pad each sequence to length-N, a number of the values of the circular convolution will clearly be equal to zero. Also, from section 5.8 above, the only effect of the zero-padding is to increase the resolution of the DFT spectra. Now, the result of the circular convolution is itself a length-N sequence. In fact, the circular convolution is equal to the length-N linear convolution in this case.

Hence, if we zero-pad $x(n)$ and $h(n)$, take their DFTs, multiply them and take the IDFT, the result is the length-N sequence $y(n)$. In this way, we can perform linear filtering using the DFT, as an alternative to time-domain convolution. At first sight, it might seem that the convolution approach is more direct and is therefore likely to be preferable. However, as we shall see in chapter 7, the existence of fast algorithms for computing the DFT – fast Fourier transform or FFT algorithms – renders the frequency-domain approach computationally more efficient overall.

5.11 SUMMARY

In this chapter, we have considered Fourier representations of a finite, length-N sequence $x^*(t) = x(n)$ corresponding to a CT signal $x(t)$ sampled with $f_s = 1/T$ and band-limited to a maximum frequency $f_{\max} = B$. The restriction to a finite length is presumed to result from truncation of the sequence for computer analysis. Signals encountered in practice are only infrequently periodic – more commonly they are quasi-periodic or aperiodic. Hence, the discrete form of the Fourier series (producing a line spectrum) is not generally appropriate.

One possible Fourier representation for DT signals is $X^*(j\omega)$. This was arrived at by, first, recognising that the train of Dirac delta pulses used to sample $x(t)$ is clearly periodic with period T and so has a Fourier series (of a rather simple form) which gives a series expression for $x^*(t)$. We then found $X^*(j\omega)$ from the Fourier transform of $x^*(t)$. This representation is periodic in frequency with period $\omega_s = 2\pi/T$. In the base-band, the spectral representation is equal to the spectrum, $X(j\omega)$, of $x(t)$. Two severe problems arise, however, with $X^*(j\omega)$. First, it assumes $x(n)$ is infinite in extent. Second, it is a continuous function of frequency. As such, it is intrinsically unsuited to computer analysis of signals and systems.

We then studied the discrete Fourier transform (DFT) which overcomes these objections. The starting point for deriving the DFT was a numerical integration of the Fourier transform using the rectangular rule, followed by a truncation to length-N and discretisation of the frequency variable ω. The frequency spacing of this discrete variable, m, was $2\pi/NT$. We showed that this corresponded to a maximally-efficient representation of the spectrum, using the minimum number of values according to the Nyquist criterion. Like $X^*(j\omega)$, the DFT spectrum is periodic with period $2\pi/T$ in hertz frequency or N in the discrete variable m. In this case, however, the spectrum in the base-band is a sampled version of $X(j\omega)$. Because the spectrum is sampled, the DFT in effect takes $x(n)$ to be periodic in time, with period $n = N$. This is a consequence of the important principle of time-frequency duality. A signal which is sampled in time is periodic in the frequency domain with period set by the sampling rate; a sampled frequency spectrum implies a signal which is similarly periodic in time. These relationships are summarised in table 5.1.

We next considered the inverse DFT, and found that this had a remarkably simi-

domain	discrete variable	sample spacing	period
time	n	T	$m = N \quad P = NT$
frequency	m	$2\pi/NT$	$n = N \quad \omega_s = 2\pi/T$

Table 5.1 Periodicity properties of DFT in time and frequency domains.

lar form to the forward DFT. Indeed, apart from a multiplying constant of $1/N$ and a change of sign of one of the discrete variables (n or m), they are identical. This similarity of form has two important consequences: one practical and the other theoretical. From a practical point of view, it becomes possible to use essentially the same algorithm for both DFT and IDFT. From a theoretical point of view, there exists a duality between time and frequency domains – as mentioned above.

In addition to the periodicity properties and time-frequency duality, the DFT possesses many other important properties. Not surprisingly, given that the DFT is the z-transform taken at equally-spaced points around the unit circle, these tend to mirror the properties of the z-transform. The linearity, shift and convolution properties are instances of this. The dual properties are also of importance in the study of discrete-time signals and systems. The DFT has important symmetry properties which again have both practical and theoretical significance. If $x(n)$ is real (as is usually the case if this sequence represents samples of a physical, real-world signal), then the magnitude spectrum has even symmetry about the Nyquist point $m = N/2$ while the phase spectrum has odd symmetry about this point. The DFT coefficient $X(0)$ is real and corresponds to the average ('dc') value of the sequence. The coefficient $X(N/2)$ is also real, and should be zero if $x(t)$ is band-limited. We showed that for the case of $x(n)$ real, the same number of values (N) is required to specify the signal in the frequency domain as in the time domain. The practical importance of this is that (given that two numbers are necessary to specify a complex number) not all coefficients have to be processed and stored in computing the DFT.

As the $X(m)$ values are generally complex, we can specify a signal by both its magnitude and phase spectrum, $|X(m)|$ and $\angle X(m)$ respectively. We then developed a Parseval relation for the DFT. Since the DFT assumes $x(n)$ to be periodic, with exactly one period contained in the length-n analysis window, we can treat $x(n)$ as a finite-power signal. We showed that the average power in the signal could be evaluated in either the time or the frequency domain. The frequency domain measure is arrived at by averaging values of $|X(m)|^2$, which is consequently called the power spectrum.

We next turned our attention to the effect on the spectrum of truncating the DT signal to blocks of finite length for analysis. Truncation amounts to multiplication of the signal in the time domain by a rectangular window. By the convolution properties of the DFT, the effect of this is to convolve the two spectra – that of the signal and that of the window – in the frequency domain. Hence, it is important to understand the spectral properties of the rectangular window and, indeed, any other form of window that we might wish to use. We found that the magnitude spectrum of the rectangular window is a form of Dirichlet function, and the phase spectrum is linear. Intriguingly, the Dirichlet function has its zeros at integer values of the frequency index m ($\neq 0$ or N). Thus, this function aliases a (periodic with period N) Kronecker delta function and convolution with it is an identity operation. To this extent, multiplication by the rectangular window 'has no effect'. This statement, however, needs careful qualification. What it really means is that the rectangular window is an implicit part of the DFT. That is, the DFT correctly yields the spectrum of the signal within the window which, crucially, is considered as one period of an infinite, period-N sequence.

However, the placement of the window when analysing a signal in blocks is essentially arbitrary. Because the DFT assumes the signal to be infinite and periodic, any discontinuities at the window ends are taken to be essential, integral features of the signal rather than (as is really the case) mere artifacts of the window shape and its placement. The spectral effect of these artifacts is called *leakage*. Hence, it is common practice to use a window (such as the Hamming window, or the Hamming taper) which introduces less severe discontinuities at window ends, so reducing (but not eliminating) leakage.

With our understanding of the spectral properties of the rectangular window, we broached the topic of interpolation by zero-padding. By taking a length-N window and appending $(N - K)$ zeros to give a length-K sequence before taking the DFT, the spacing of the spectral samples is reduced from $2\pi/NT$ to $2\pi/KT$. The zero-padding process was introduced as equivalent to multiplication in the time domain by a rectangular window of slightly different periodicity from that used for analysis of a signal into blocks. Again, the magnitude spectrum is a Dirichlet function but – for the representative case of $K = 2N$ – the zeros are at even values of the frequency index m ($\neq 0$ or N) rather than at all integer values. Thus, when convolving the signal spectrum with this function, the non-zero values at odd m interpolate between the existing spectral samples. By this means, we can improve (for instance) our estimate of the location of a spectral peak. Further, by time-frequency duality, we can apply similar processing to interpolate the time-domain signal. Thus, if we take the DFT, zero-pad, and take the inverse, we achieve an enhanced resolution in the time-domain.

The study of interpolation allowed us to consider further a question posed in chapter 1, namely how to reconstruct a CT signal from its samples. This is ideal interpolation or digital-to-analogue conversion. Since the Fourier representation $X^*(j\omega)$ in the baseband is a scaled version of $X(j\omega)$, we can recover $x(t)$ from $x^*(t)$ by filtering with an ideal low-pass filter. Such a filter has a rectangular shape in the frequency domain, multiplying $X^*(j\omega)$. Hence, we must convolve the sample values $x^*(t)$ in the time domain by the IDFT of the rectangular filter shape, which is a sinc function. This sinc function is zero at values of t which coincide with the sampled values (kT) except at $k = n$ when the function has unit height and yields the value $x(n)$ in the convolution.

By the time-convolution property, the DFT allows us to process a DT signal $x(n)$ by a linear system, or filter, having impulse response $h(n)$ to produce an output $y(n)$ by multiplication in the frequency domain. Zero-padding of $x(n)$ and $h(n)$ is necessary beforehand to account for the fact that the (linear) convolution of two sequences of lengths J and K has length $(J + K - 1)$. The question then arises: is frequency-domain multiplication any more efficient than time-domain convolution, given that the former will require us to find two DFTs and one IDFT? In chapter 7, however, we will see that the DFT can be computed via one of a class of efficient algorithms known collectively as the fast Fourier transform (FFT). The existence of class of FFT algorithms dramatically increases the attractiveness of the DFT approach to linear filtering.

5.12 Exercises

5.1 As a general rule, signals that are extensive in the time domain (e.g. a sine wave) are compact in the frequency domain and, as a result of time-frequency duality, *vice versa*. Can you think of any counter-examples (e.g. signals which are extensive in both time and frequency)?

5.2 Find the DFT spectra (magnitude and phase) of the following sequences:

(i) $x(n) = 5\cos(6\pi n/N)$

(ii) $x(n) = \sin(\sqrt{15}\pi n/N)$

(iii) $x(n) = \sin(2\pi n/N) + 0.5\cos(\frac{2\pi n}{N} + \pi/4)$

What difficulties follow in (ii) from the fact that $\sqrt{15}$ is irrational?

5.3 Find the DFT spectrum of the signal $x(t) = \sin(2\pi f_0 t + \pi/8)$ when sampled at 10 times the Nyquist rate and for an analysis window of length N into which exactly 8 periods of the signal fit.

5.4 Find the DFT (magnitude and phase) of the ramp function with a discontinuity at the centre of its (length-8) window:

$$x(n) = \begin{cases} n & 0 \le n \le 3 \\ (n-7) & 4 \le n \le 7 \end{cases}$$

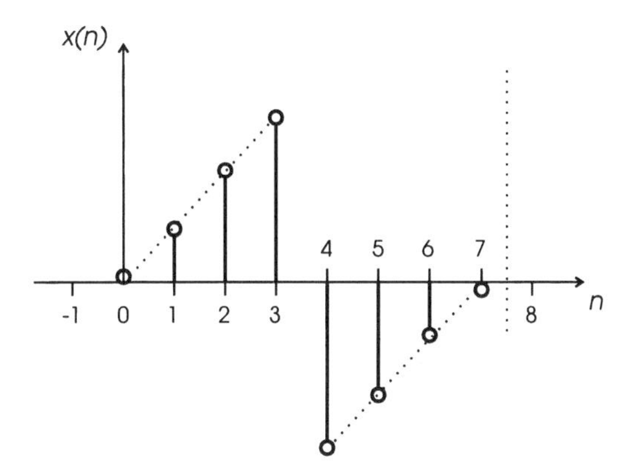

Fig. 5.13 Ramp function with discontinuity at centre of window.

as depicted in figure 5.13. (Note that you will need to know how to evaluate $\sum_{n=0}^{N-1} nq^n$. Recognising that $nq^n = q\frac{d}{dq}q^n$, this sum can be found from the familiar summation formula for $\sum_{n=0}^{N-1} q^n$.)

5.5 Using l'Hôpital's rule, confirm the result in example 6 that:

$$x(0) = \mathcal{D}^{-1}[X(m) = 1]\big|_{n=0} = 1$$

The rule can be stated for the general case of complex z as follows. If $f(z)$ and $g(z)$ are both analytic at $z = a$, but:

$$\begin{aligned} f(a) &= f'(a) = \ldots = f^{(k)}(a) = 0 \\ \text{and} \quad g(a) &= g'(a) = \ldots = g^{(k)}(a) = 0 \end{aligned}$$

whereas $f^{(k+1)}(a)$ and $g^{(k+1)}(a)$ are not both equal to zero, then:

$$\lim_{z \to a} \frac{f(a)}{g(a)} = \frac{f^{(k+1)}(a)}{g^{(k+1)}(a)}$$

The form of this for real z is obvious.

5.6 In section 5.5.4, the modulation property of the DFT was presented as the dual of the shift property. Derive the modulation property from first principles (i.e. by substitution into the DFT defining equation). Find the DFT spectrum of $\cos(2\pi n/N)$ modulated by $\cos(2\pi kn/N)$:

(i) from first principles;

(ii) using the modulation property;

(iii) using the appropriate convolution property.

5.7 Derive the modulation property from the frequency-convolution property:

$$\mathcal{D}[x(n)y(n)] = \frac{1}{N} X(m) * Y(m)$$

5.8 What is the relation (if any) between circular convolution as met in this chapter, and circular convolution as met in exercise 3.3?

5.9 Because the DFT corresponds to the evaluation of the z-transform around the unit circle, the two transforms share many common properties. Does the ZT possess analogous *uniqueness* and *duality* properties to the DFT?

5.10 From the symmetry properties of the DFT, show that (apart from a constant scaling factor, k) the IDFT can be found using the forward DFT by time-reversing the spectral sequence. That is:

$$\begin{aligned} \text{if} \quad \mathcal{D}^{-1}[X(m)] &= x(n) \\ \text{then} \quad \mathcal{D}[X(-m)] &= kx(n) \end{aligned}$$

What is the value of the scaling factor?

5.11 Find the DFT spectrum of $x(n) = \cos(1.5\pi n)$. By taking the IDFT, show that this sequence aliases a lower-frequency cosine wave. What is this lower, aliasing frequency?

5.12 Find the DFT spectrum of the Hamming window, and compare it with that of the rectangular window.

5.13 In example 5 of chapter 3, it was found that the z-transform for a sampled sine wave did not converge on the unit circle. Yet the DFT, which considers signals to be composed of summed sinusoids and cosinusoids, evaluates the ZT at N points equally spaced around the unit circle. Resolve this apparent ambiguity.

6

Finite impulse response (FIR) filters

In chapter 2, we saw that the general difference equation (2.5) for a DT system is:

$$y_n = \sum_{r=0}^{M} b_r x_{n-r} - \sum_{k=1}^{N} a_k y_{n-k}$$

If the a_k coefficients are all identically zero:

$$y_n = \sum_{r=0}^{M} b_r x_{n-r}$$

Hence, the output depends only upon the input. That is, there is no feedback and the system poles appear only at the z-plane origin. As we have seen earlier, such systems have a finite impulse response (FIR) whose sequence values are identical to the b_r's. Thus, the length of this sequence is $(M + 1)$. From this point on, to conform with the literature on FIR filter design, we denote the length of the impulse response sequence, $|h(n)|$, as N.

In this chapter, we consider the properties of such FIR filters before describing two useful design techniques: Fourier series truncation and frequency sampling. The latter uses the discrete Fourier transform (DFT) which we studied in some depth in chapter 5. Because both methods rely on making an approximation to a Fourier representation, oscillatory (Gibbs) phenomena contaminate the filter's frequency response and need to be considered.

6.1 PROPERTIES OF FIR FILTERS

FIR systems have a number of important properties, of which the most important are the *stability* and *linear-phase* properties.

6.1.1 Stability of FIR filters

Since there are no feedback terms (all a_k's zero), an FIR system has no poles. As there are no poles, the stability criterion of no poles within the z-plane unit circle is unconditionally satisfied.

Hence, the design process for FIR filters is simplified relative to that for IIR filters (chapter 4) because there is no need to consider the stability of the realisation.

6.1.2 Linear-phase property

From chapter 3 (section 3.8.2), the phase response, $\phi()$, of a DT system is given by:

$$\phi(\omega T) = \angle H\left(e^{j\omega T}\right)$$

Changing the frequency variable to $\nu = \omega T$:

$$\phi(\nu) = \angle H\left(e^{j\nu}\right)$$

The negative derivative of $\phi()$ with respect to frequency is the well-known time (or group) delay function:

$$T(\nu) = -\frac{d\phi(\nu)}{d\nu} \tag{6.1}$$

If $\phi(\nu)$ is a linear function of frequency, then:

$$\phi(\nu) = -k\nu \qquad (k \text{ is a positive constant})$$

$$\text{and} \qquad T(\nu) = k \tag{6.2}$$

Since $T(\nu)$ is a constant, all sinusoidal components in the input will be delayed by the same amount (see exercise (6.1)). Hence, a constant time delay or linear-phase system will avoid phase distortion of its output. This means that the only alteration to the time-waveshape of the input is the attenuation of frequency components outside the passband. In this sense, the output waveform is an undistorted version of the input.

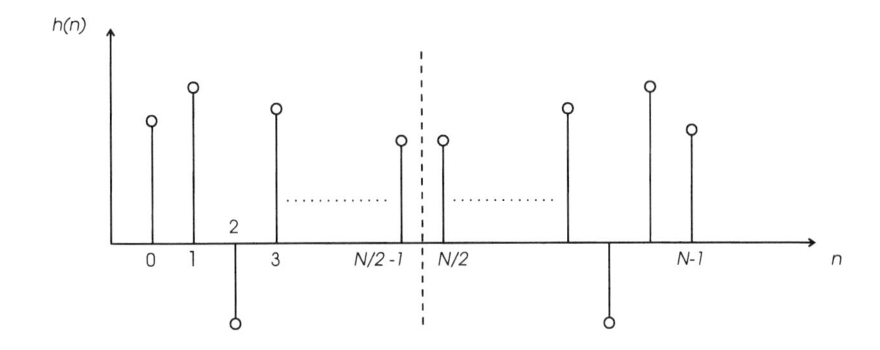

Fig. 6.1 If $N = |h(n)|$ is even, the impulse response of an FIR system will have an axis of symmetry about a mid-point between two sequence values.

The IIR filter design techniques of chapter 4 paid no attention to the phase response, and gave no guarantee of linear phase (rather the opposite, in fact). One of the major advantages of FIR filters is that they can be designed to have linear phase.

We show now that the necessary condition for an FIR filter to have linear phase is that its impulse response is symmetric:

$$h(n) = h(N - 1 - n)$$

where, as stated above, N is the sequence length, $|h(n)|$, of the impulse response.

To see why this is so, consider two mutually-exclusive and exhaustive cases.

Case 1: N is even

In this situation, $h(n)$ will have an axis of symmetry at the mid-point of the sequence, which will fall between two consecutive values, as depicted in figure 6.1.

Applying the z-transform:

$$H(z) = \sum_{n=0}^{N-1} h(n)z^{-n}$$

We can split $h(n)$ in two equal-length sub-sequences:

$$H(z) = \sum_{n=0}^{\frac{N}{2}-1} h(n)z^{-n} + \sum_{n=\frac{N}{2}}^{N-1} h(n)z^{-n} \tag{6.3}$$

As n is a 'dummy' variable, it can be replaced by $(N - 1 - n)$ in the second summation:

$$
\begin{aligned}
H(z) &= \sum_{n=0}^{\frac{N}{2}-1} h(n)z^{-n} + \sum_{N-1-n=\frac{N}{2}}^{N-1-n=N-1} h(N-1-n)z^{-(N-1-n)} \\
&= \sum_{n=0}^{\frac{N}{2}-1} h(n)z^{-n} + \sum_{n=\frac{N}{2}-1}^{n=0} h(N-1-n)z^{-(N-1-n)} \\
&= \sum_{n=0}^{\frac{N}{2}-1} h(n)z^{-n} + \sum_{n=0}^{\frac{N}{2}-1} h(N-1-n)z^{-(N-1-n)}
\end{aligned}
$$

But the symmetry property demands $h(n) = h(N-1-n)$, so that:

$$
H(z) = \sum_{n=0}^{\frac{N}{2}-1} h(n)\left[z^{-n} + z^{-(N-1-n)}\right]
$$

Replacing z by $e^{j\omega T}$ enables us to determine the frequency response of the system:

$$
\begin{aligned}
H\left(e^{j\omega T}\right) &= \sum_{n=0}^{\frac{N}{2}-1} h(n)\left[e^{-j\omega nT} + e^{-j\omega(N-1-n)T}\right] \\
&= e^{-j\omega\left(\frac{N-1}{2}\right)T} \sum_{n=0}^{\frac{N}{2}-1} h(n)\left[e^{-j\omega\left(n-\frac{N-1}{2}\right)T} + e^{-j\omega\left(\frac{N-1}{2}-n\right)T}\right]
\end{aligned}
$$

Since $\cos\theta = \frac{1}{2}\left(e^{j\theta} + e^{-j\theta}\right)$:

$$
H(e^{j\omega T}) = e^{-j\omega\left(\frac{N-1}{2}\right)T} \sum_{n=0}^{\frac{N}{2}-1} 2h(n)\cos\omega\left(n - \frac{N-1}{2}\right)T \qquad (6.4)
$$

The summation, S say, is purely real, but may be positive or negative. Hence, we can express equation (6.4) in polar form as:

$$
H\left(e^{j\omega T}\right) = Se^{-j\omega\left(\frac{N-1}{2}\right)T}
$$

Thus, with $\nu = \omega T$, the phase function is:

$$
\phi(\nu) = \begin{cases} -\left(\frac{N-1}{2}\right)\nu & S > 0 \\ -\left(\frac{N-1}{2}\right)\nu - \pi & S < 0 \end{cases}
$$

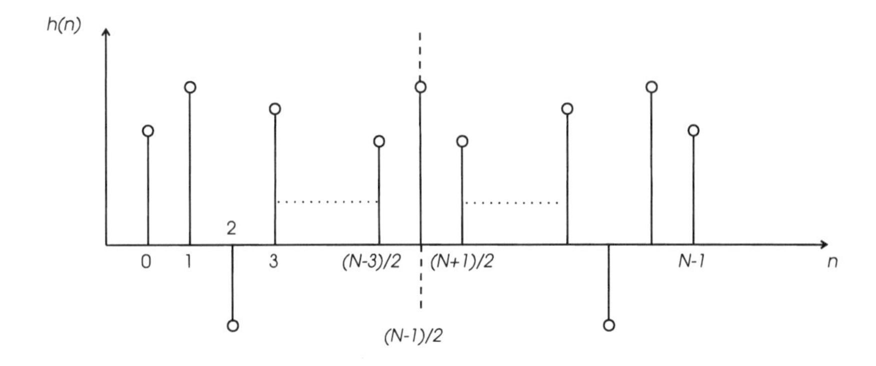

Fig. 6.2 If $N = |h(n)|$ is odd, the impulse response will have an axis of symmetry which corresponds to the central sequence value.

From equation (6.1) and noting that π is a constant (independent of ν), this implies a constant time delay of $(N-1)/2$ samples for all frequency components in the input. Since N is even, the delay corresponds to a non-integer number of samples.

Case 2: N is odd

In this case, $h(n)$ will have an axis of symmetry which corresponds to the value at the centre of the sequence, as depicted in figure 6.2.

The earlier equation (6.3) for the system function in the N-even case is modified to:

$$H(z) = h\left(\frac{N-1}{2}\right) z^{-\frac{(N-1)}{2}} + \sum_{n=0}^{\left(\frac{N-1}{2}\right)-1} h(n) z^{-n} + \sum_{n=\left(\frac{N-1}{2}\right)+1}^{N-1} h(n) z^{-n}$$

Proceeding as before, we find:

$$H\left(e^{j\omega T}\right) = e^{-j\omega\left(\frac{N-1}{2}\right)T}\left[h\left(\frac{N-1}{2}\right) + \sum_{n=0}^{\frac{N-3}{2}} 2h(n)\cos\omega\left(n - \frac{N-1}{2}\right)T\right]$$

Again, the expression in the square braces is purely real and, as before:

$$\phi(\nu) = \begin{cases} -\left(\frac{N-1}{2}\right)\nu & S > 0 \\ -\left(\frac{N-1}{2}\right)\nu - \pi & S < 0 \end{cases}$$

Thus, there is a constant time delay of $(N-1)/2$ samples for all frequencies. In this case (N odd), the delay is by an integer number of samples.

So, since the possibilities of N even and N odd are exhaustive, cases 1 and 2 together confirm that phase distortion is avoided in an FIR filter by ensuring a symmetric impulse response.

6.2 FOURIER SERIES TRUNCATION

Our purpose in realising an FIR filter is to find a DT system whose system function $H(z)$ approximates a desired magnitude response $|H_d(e^{j\omega T})|$ and which has the linear-phase property.

A real function of time, $x(t)$, periodic with period P can be represented by the Fourier series (equation (1.18)):

$$x(t) = \sum_{-\infty}^{\infty} c_n e^{\frac{2\pi jnt}{P}}$$

We are by now familiar with the fact that the magnitude response of a discrete-time system, $|H_d(e^{j\omega T})|$, is periodic in frequency with period determined by the sampling rate, $\omega_s = 2\pi/T$ – as depicted in figure 6.3. Hence, we can express the magnitude response as a Fourier series where the basis functions are functions of frequency rather than time. Using $\nu = \omega T$ as a frequency variable, $H_d(\nu)$ has period $\nu_s = 2\pi$. So, replacing the function $x()$ of time in the Fourier series above by a function $|H_d()|$ of the frequency variable ν with period $P = 2\pi$:

$$|H_d(\nu)| = \sum_{-\infty}^{\infty} c_n e^{jn\nu}$$

$$\text{and} \quad \left|H_d\left(e^{j\omega T}\right)\right| = \sum_{-\infty}^{\infty} c_n e^{jn\omega T}$$

This Fourier series representation of the magnitude response has, of course, an infinite number of coefficients, c_n. So we truncate the series to its N lowest-order terms, with the c_n coefficients ($|n| \leq Q$, say) specifying the (finite) impulse response. These are evaluated with the Fourier analysis equation (section 1.9.1) as:

$$c_n = \frac{1}{2\pi} \int_{-\pi}^{\pi} |H_d(\nu)|\, e^{-jn\nu} d\nu \tag{6.5}$$

In general, the c_n coefficients are complex. However, $|H_d(\nu)|$ is an even function of frequency, so here the coefficients are real and the Fourier series has cosine components only (see exercise 1.8). Thus, equation (6.5) simplifies to:

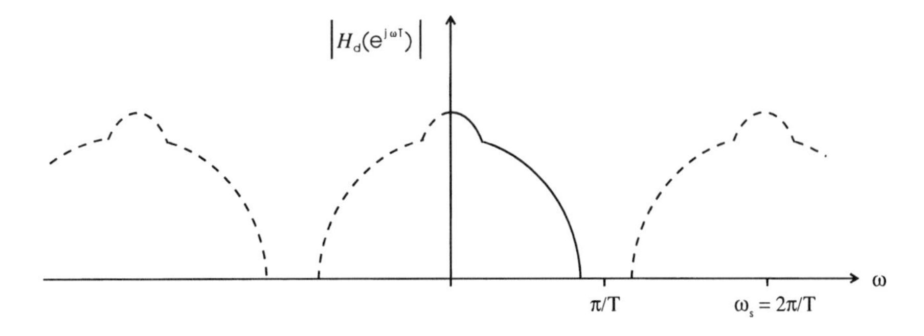

Fig. 6.3 The magnitude response $|H_d(e^{j\omega T})|$ of a DT system is periodic in frequency with period determined by the sampling rate, $\omega_s = 2\pi/T$.

$$c_n = \frac{1}{\pi}\int_0^{\pi} |H_d(\nu)| \cos(n\nu) d\nu \qquad n \geq 0 \tag{6.6}$$

and
$$c_{-n} = c_n \tag{6.7}$$

As stated above, we must truncate the infinite number of c_n's to a finite number of coefficients. Note that this number, $(2Q + 1)$, must be odd as a consequence of the symmetry in equation (6.7).

This truncation leads to an approximation (reverting to $e^{j\omega T}$ as a frequency variable):

$$\begin{aligned}\hat{H}_d(e^{j\omega T}) &\approx H_d(e^{j\omega T}) \\ &= \sum_{n=-Q}^{Q} c_n e^{jn\omega T}\end{aligned}$$

In terms of z ($= e^{j\omega T}$ for evaluation around the unit circle):

$$\hat{H}_d(z) = \sum_{n=-Q}^{Q} c_n z^n$$

This is, of course, the system function of an FIR filter whose impulse response is:

$$\{c_{-Q}, c_{-Q+1} \ldots, c_0, c_1, \ldots, c_{Q-1}, c_Q\}$$

The filter is, however, non-causal because of the positive powers, n, of z which lead to the Q terms in the impulse response appearing before c_0 at time index 0. To overcome this problem, we can delay the output by Q samples. By the right-shift property, this corresponds to multiplying by z^{-Q}, i.e. placing Q poles at the z-plane origin. Then:

$$\begin{aligned} H(z) &= z^{-Q}\hat{H}_d(z) \\ &= z^{-Q}\sum_{n=-Q}^{Q} c_n z^n \\ &= c_{-Q}z^{-2Q} + c_{-Q+1}z^{-2Q+1} + \\ &\quad \cdots + c_0 z^{-Q} \cdots + c_{Q-1}z^{-1} + c_Q \end{aligned}$$

If we make:

$$b_r = c_{Q-r} \tag{6.8}$$

i.e. let $b_0 = c_Q, b_1 = c_{Q-1}, \ldots, b_{2Q} = c_{-Q}$, then it is apparent that:

$$b_i = b_{2Q-i}$$

and the impulse response is symmetric leading to linear phase. It has duration given by the number of coefficients ($N = 2Q + 1$) minus 1, times the sampling period. That is, the impulse response duration is equal to 2QT.

The system function is:

$$H(z) = \sum_{i=0}^{2Q} b_i z^{-i}$$

Now, since:

$$\begin{aligned} H(z) &= z^{-Q}\hat{H}_d(z) \\ H(e^{j\omega T}) &= e^{-jQ\omega T}\hat{H}_d\left(e^{j\omega T}\right) \\ \text{and} \quad \left|H\left(e^{j\omega T}\right)\right| &= \left|\hat{H}_d\left(e^{j\omega T}\right)\right| \end{aligned}$$

Hence, just as we saw in chapter 3, the addition of the extra Q poles has no effect on the obtained magnitude response, which approximates the desired $|\hat{H}_d(e^{j\omega T})|$ response

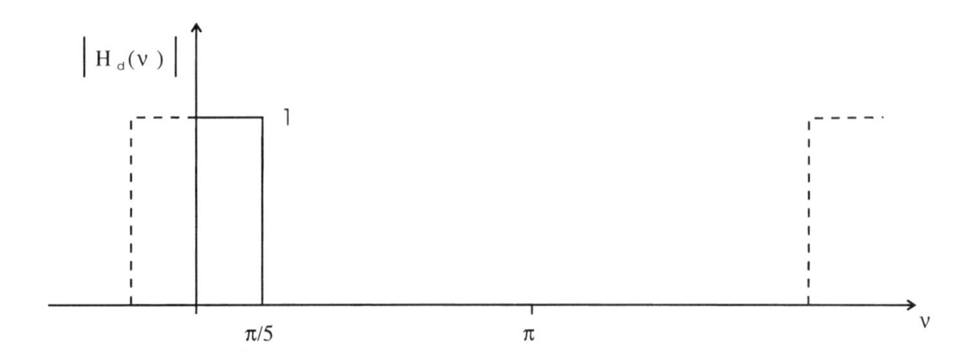

Fig. 6.4 Desired magnitude response for (ideal) low-pass FIR filter to be designed by the Fourier series truncation method.

with closeness depending upon the choice of Q. The larger is Q, the better is the approximation.

We now illustrate the use of the Fourier series method with an example.

Example 1: Design a low-pass FIR filter to have unity gain from 0 up to 400 Hz, and zero gain above 400 Hz. The sampling frequency, f_s, is 4 kHz, and the impulse response is limited to a duration of 4 ms. What is the time delay imposed on the filtered input?

Since the sampling frequency is 4 kHz, the Nyquist frequency f_N will be 2 kHz and $T = \frac{1}{4000} = 0.25$ ms. The impulse response duration is limited to 4 ms, so that:

$$\begin{aligned} 2QT &\leq 2Q(0.25 \times 10^{-3}) = 4 \times 10^{-3} \\ \text{and} \qquad Q &\leq 8 \end{aligned}$$

We choose, therefore, to implement the filter with 17 coefficients (i.e. $Q = 8$). Of course, we cannot expect to achieve anything like an ideal response – flat in the passband and with infinite attenuation in the stopband – with such a small number of coefficients. We will examine the effects of this truncation on the obtained magnitude response shortly.

The required filter coefficients are obtained from equation (6.6), with $|H_d(\nu)|$ set equal to the idealised values of 1 in the passband and 0 elsewhere. The upper limit for the integration – corresponding to the edge of the passband – is found as follows. The Nyquist frequency (2 kHz) corresponds to ν $(= \omega T) = \pi$ (i.e. half the period, ω_s) in equation (6.6). Hence, the 400 Hz break frequency corresponds to $\frac{400\pi}{2000} = \pi/5$, as depicted in figure 6.4. Thus:

$$\begin{aligned} c_n &= \frac{1}{\pi}\int_0^{\frac{\pi}{5}} \cos(n\nu)d\nu \\ &= \frac{1}{\pi}\left[\frac{\sin(n\nu)}{n}\right]_0^{\frac{\pi}{5}} \\ &= \frac{\sin\left(\frac{n\pi}{5}\right)}{n\pi} \end{aligned} \quad (6.9)$$

As expected, the coefficients are sinc functions. Evaluating these for $n = 0 \ldots Q$:

$$\begin{aligned} c_0 &= 0.20000 & c_1 &= 0.18710 \\ c_2 &= 0.15137 & c_3 &= 0.10091 \\ c_4 &= 0.04678 & c_5 &= 0 \\ c_6 &= -0.0312 & c_7 &= -0.0432 \\ c_8 &= -0.0378 \end{aligned}$$

from which the b_r coefficients can be found using equations (6.7) and (6.8).

As previously mentioned, the slight difficulty in evaluating c_0 above (in that substituting $n = 0$ directly into equation (6.9) gives zero divided by zero) is easily overcome by noting that $\sin(x) \approx x$ for x small, so that:

$$\frac{\sin\left(\frac{n\pi}{5}\right)}{n\pi} \approx \frac{\left(\frac{n\pi}{5}\right)}{n\pi} = \frac{1}{5}$$

From section 6.1.2 above, the (constant) time delay imposed on the input is:

$$\Delta t = \frac{(N-1)T}{2}$$

for a linear-phase FIR filter, or half the length of the impulse response. Hence, the delay is:

$$\Delta t = \frac{(17-1)0.25}{2}\ \text{ms} = 2\ \text{ms}$$

i.e. half of 4 ms.

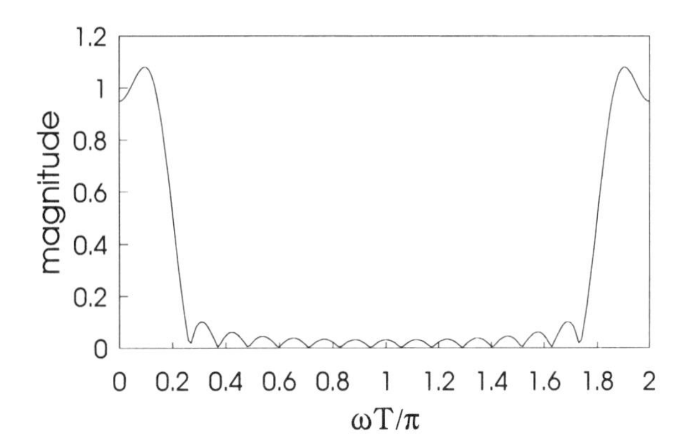

Fig. 6.5 Obtained magnitude response of low-pass FIR filter with cut-off frequency of 400 Hz ($\omega T/\pi = 0.2$) designed by FS truncation with 17 coefficients.

6.3 WINDOWING

In the Fourier series truncation method above, we obtained an infinite impulse response sequence corresponding to the desired frequency response, $H_d(e^{j\omega T})$, and truncated it. For our example design, the desired magnitude response had ideally infinite fall-off at the break frequency $\nu = \pi/5$, yet we abruptly truncated the impulse response sequence to just 17 values. Consequently, it is reasonable to ask what effect this truncation has on the actual magnitude response.

Figure 6.5 shows the magnitude response for the obtained filter. As a result of the truncation to length $N = 17$, we see that the fall-off is far from infinite at the break frequency and there is unwanted ripple in both the pass- and stopbands. From earlier study of the Fourier series, you may recognise this as the *Gibbs phenomenon*.

We can understand how this arises using the notion of windowing introduced in section 5.7. Expressing our truncated impulse response $h(n)$ as the product of the desired (infinite) impulse response $h_d(n)$ and a rectangular window function:

$$h(n) = h_d(n)w(n) \tag{6.10}$$

where the window function is:

$$w(n) = \begin{cases} 1 & 0 \leq n \leq (N-1) \\ 0 & \text{otherwise} \end{cases}$$

Taking z-transforms of equation (6.10) using the convolution property:

$$H(z) = H_d(z) * W(z)$$

$$\text{with} \quad W(z) = \sum_{n=0}^{N-1} z^{-n}$$

$$\text{so that} \quad W(e^{j\omega T}) = \sum_{n=0}^{N-1} e^{-j\omega n T}$$

Again, using the fact that this is a truncated geometric series, we find the spectrum of the rectangular window to be:

$$W(e^{j\omega T}) = e^{-j\omega((N-1)/2)T} \left(\frac{\sin(\omega N T/2)}{\sin(\omega T/2)} \right)$$

Note that this result could have been obtained from equation (5.15) for $W(m)$ (the spectrum of a rectangular window in the discrete frequency domain) of the previous chapter by substituting the continuous variable $\omega N T/2\pi$ for the frequency index m.

It is easily seen that the phase is linear, corresponding to a delay of $(N-1)/2$ samples.

The magnitude is given by:

$$\left| W\left(e^{j\omega T}\right) \right| = \left| \frac{\sin(\omega N T/2)}{\sin(\omega T/2)} \right| \tag{6.11}$$

Figure 6.6 shows this function for the case $N = 17$. There is a main lobe between $\omega T/\pi = 2/N$ or approximately 0.118 rad.s^{-1} in this case (although only the positive part is shown in the figure). However, there are also side lobes. It is these side lobes which cause the problem of ripple, since this is the function with which our desired frequency response is convolved. Ideally, we would like it to be infinitely narrow (recall that convolution with the delta function is an identity operation) so as to give a sharp transition at the break points with no side lobes. As N increases, the width of the main lobe does indeed decrease as desired but, unfortunately, the amplitudes of the side lobes do not. They remain a fixed proportion (some 22.5% for the first side lobe, or -13 dB down) of the main lobe amplitude, N.

Because of this problem with the rectangular window, many alternative window functions have been devised. These taper the window smoothly to zero at its ends as a way of reducing the heights of the side lobes. Unfortunately, however, this widens the main lobe and so leads to a less sharp transition at the break points. We have already met

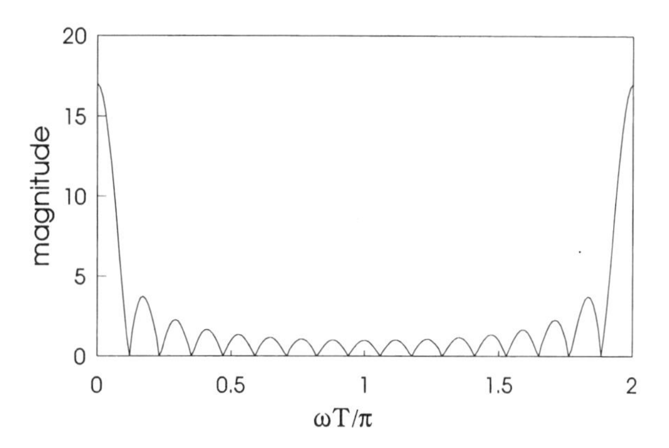

Fig. 6.6 Magnitude spectrum of rectangular window for $N = 17$.

the Hamming window, which is often used in this way, in the previous chapter (equation (5.14)).

Some other well-known window functions are:

Hanning: $w(n) = \frac{1}{2}\left[1 - \cos\left(\frac{2\pi n}{N-1}\right)\right] \qquad 0 \le n \le N-1$

Blackman: $w(n) = 0.42 - 0.5\cos\left(\frac{2\pi n}{N-1}\right) + 0.08\cos\left(\frac{4\pi n}{N-1}\right) \qquad 0 \le n \le N-1$

Bartlett: $w(n) = \frac{2n}{N-1} \qquad 0 \le n \le \frac{N-1}{2}$

$w(n) = 2 - \frac{2n}{N-1} \qquad \frac{N-1}{2} \le n \le N-1$

Of course, truncation to just $N = 17$ filter coefficients is rather unrealistic. It has been done here to keep the mathematics tractable so that general principles can be emphasised. The problems of slow roll-off and ripple (Gibbs phenomenon) can be reduced by increasing N but this obviously increases the complexity of the filter.

6.4 FREQUENCY SAMPLING METHOD

Let the impulse response of the discrete-time FIR filter to be designed be $h(n)$, with $0 \le n \le (N-1)$. The fact that the frequency response of a DT system is formally equivalent to the discrete Fourier transform of $h(n)$ can be used as the basis of a design technique for FIR filters. Because the DFT is discrete in frequency, i.e. it samples

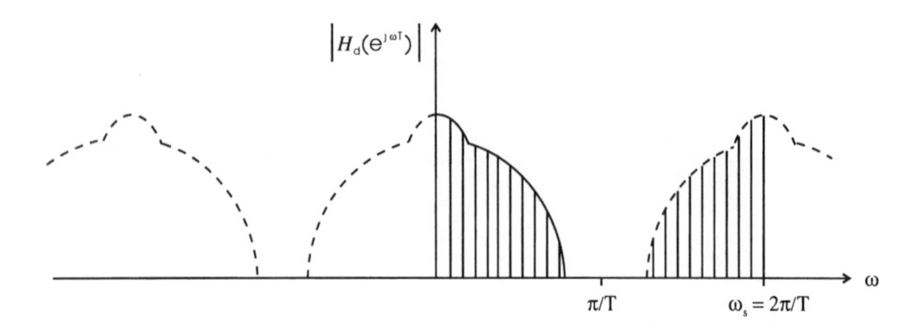

Fig. 6.7 In the frequency sampling method of FIR filter design, the desired frequency response is sampled to yield a sequence $H(m)$. For ease of drawing, this is depicted here in terms of the magnitude response $|H_d(e^{j\omega T})|$ for the case of N odd.

the spectrum of the corresponding CT function ($h(t)$ here), we call this the frequency sampling method.

In the FS truncation method, we consider the $h(n)$ values (identical to the b_r filter coefficients) to be the Fourier series coefficients of the desired frequency response, which is a continuous function of frequency. In the frequency sampling method, however, we consider the $h(n)$ values to be the DFT coefficients of the desired frequency response, which has been sampled to yield a (complex) sequence $H(m)$. This is depicted in figure 6.7 in terms of the real part of $H(m)$, i.e. the magnitude response.

Limiting the impulse response, $h(n)$, of a DT system to be a length-N, right-sided sequence, we can write its DFT from equation (5.4) of the previous chapter as:

$$H(m) = \sum_{n=0}^{N-1} h(n)e^{-2\pi jmn/N} \tag{6.12}$$

where the $H(m)$'s are the values of the system's frequency response, sampled in the frequency domain. The DFT representation treats the length-N analysis window as if it were a single period. Consistent with figure 6.7, we will take this period to extend from zero frequency up to the sampling frequency. In general, $H(m)$ will be complex but we require $h(n)$ to be real.

So, the frequency sampling design procedure consists of finding real $h(n)$ values such that the $H(m)$'s in equation (6.12) correspond to the desired frequency response for $0 \leq m \leq (N-1)$, i.e. for $0 \leq \omega \leq 2\pi/T$. Also, to ensure linear phase, the $h(n)$ sequence must be symmetric, as well as real-valued.

Applying the inverse DFT (equation (5.8)) to equation (6.12) above gives:

$$h(n) = \frac{1}{N}\sum_{m=0}^{N-1} H(m)e^{2\pi jmn/N} \tag{6.13}$$

Now, $H(z)$ for the designed filter is:

$$H(z) = \sum_{n=0}^{N-1} h(n)z^{-n} \tag{6.14}$$

Substituting (6.13) into (6.14) and reversing the order of summation gives:

$$H(z) = \frac{1}{N}\sum_{m=0}^{N-1} H(m)\left[\sum_{n=0}^{N-1}\left(e^{\frac{2\pi jm}{N}}z^{-1}\right)^n\right]$$

The term in square brackets is a finite, or truncated, geometric series with common ratio $(e^{\frac{2\pi jm}{N}}z^{-1})$. Since the common ratio will not in general equal 1 for a general value of z, this can be evaluated in the usual way as:

$$\begin{aligned}\sum_{n=0}^{N-1}\left(e^{\frac{2\pi jm}{N}}z^{-1}\right)^n &= \frac{1-\left(e^{\frac{2\pi jm}{N}}z^{-1}\right)^N}{1-\left(e^{\frac{2\pi jm}{N}}z^{-1}\right)} \\ &= \frac{1-z^{-N}}{1-\left(e^{\frac{2\pi jm}{N}}z^{-1}\right)}\end{aligned}$$

and, hence:

$$H(z) = \frac{1}{N}\sum_{m=0}^{N-1} H(m)\frac{1-z^{-N}}{1-\left(e^{\frac{2\pi jm}{N}}z^{-1}\right)}$$

The approximation to the desired system function is then found by putting $z = e^{j\omega T}$:

$$\begin{aligned}\hat{H}_d(e^{j\omega T}) &= \frac{1}{N}\sum_{m=0}^{N-1} H(m)\frac{1-e^{-j\omega NT}}{1-\left(e^{\frac{2\pi jm}{N}}e^{-j\omega T}\right)} \\ &= \frac{1}{N}\sum_{m=0}^{N-1} H(m)\frac{1-e^{-jN\left(\omega T-\frac{2\pi m}{N}\right)}}{1-\left(e^{\frac{2\pi jm}{N}}e^{-j\omega T}\right)} \qquad \text{(since } e^{2\pi jm}=1\text{)}\end{aligned}$$

Some straightforward algebraic manipulation results in:

$$\hat{H}_d\left(e^{j\omega T}\right) = e^{-j\frac{(N-1)\omega T}{2}} \sum_{m=0}^{N-1} H(m) e^{j\frac{(N-1)m\pi}{N}} \frac{\sin\left[N\left(\omega T - \frac{2\pi m}{N}\right)/2\right]}{N \sin\left[\left(\omega T - \frac{2\pi m}{N}\right)/2\right]} \tag{6.15}$$

Let us evaluate equation (6.15) at the sample points $\omega T = 2\pi m/N$ in the frequency domain, $0 \leq m \leq (N-1)$. Since we are dealing with fixed values of m, we can drop the summation. Then, using the small-angle approximation $\sin(x) \approx x$ for small (actually zero!) x:

$$\begin{aligned} \hat{H}_d\left(\frac{2\pi m}{N}\right) &= e^{-j\frac{(N-1)\pi m}{N}} H(m) e^{j\frac{(N-1)m\pi}{N}} \frac{\left[N\left(\omega T - \frac{2\pi m}{N}\right)/2\right]}{N\left[\left(\omega T - \frac{2\pi m}{N}\right)/2\right]} \\ &= H(m) \end{aligned}$$

Thus, the method does indeed yield the desired frequency response at the sample points. At other points, the $\sin(Nx)/\sin(x)$ Dirichlet term in equation (6.15) acts as an interpolation function.

If we now choose the $H(m)$'s so that the summation in equation (6.15) is purely real, call it S, then $\hat{H}_d(e^{j\omega T})$ will have the linear-phase property since::

$$\hat{H}_d\left(e^{j\omega T}\right) = Se^{-j\frac{(N-1)\omega T}{2}} = Se^{j\phi}$$

In this case, for both possible polarities of S, the group delay function is:

$$-\frac{d\phi}{d(\omega T)} = \frac{(N-1)}{2}$$

giving a constant delay of $\frac{(N-1)}{2}$ samples.

To achieve this, we require $H(m)e^{j\frac{(N-1)m\pi}{N}}$ purely real. That is:

$$\begin{aligned} H(m) &= \pm|H(m)|e^{j\psi_m} & (6.16) \\ \text{with } \psi_m &= -\frac{(N-1)m\pi}{N} & (6.17) \end{aligned}$$

It remains to ensure that the $h(n)$'s are real. From the IDFT, equation (6.13):

$$h(n) = \frac{1}{N} \sum_{m=0}^{N-1} H(m) e^{2\pi jmn/N}$$

Consider the summand, $S(m)$ say, and the effect of substituting $(N - m)$ for m in the summand:

$$\begin{aligned} S(m) &= H(m)e^{2\pi jmn/N} \\ S(N-m) &= H(N-m)e^{2\pi j(N-m)n/N} \\ &= H(N-m)e^{-2\pi jmn/N} \end{aligned}$$

In line with our earlier study of the symmetry properties of the DFT (section 5.5.6), these are complex conjugates such that:

$$H(m) = \tilde{H}(N-m) \tag{6.18}$$

Note that this condition can only hold for $m \geq 1$. In particular, it cannot be satisfied for $m = 0$ since $H(N)$ is undefined. The upper limit on m is $(N/2) - 1$ for N even and $(N-1)/2$ for N odd.

Suppose we impose equation (6.18) as a condition. Substituting into (6.13):

$$h(n) = \begin{cases} \frac{1}{N}\left[H(0) + \sum_{m=1}^{(N/2)-1} 2Re\left\{H(m)e^{2\pi jmn/N}\right\} - H\left(\frac{N}{2}\right)\right] & N \text{ even} \\ \frac{1}{N}\left[H(0) + \sum_{m=1}^{(N-1)/2} 2Re\left\{H(m)e^{2\pi jmn/N}\right\}\right] & N \text{ odd} \end{cases} \tag{6.19}$$

Now, since $\psi_0 = 0$ in equation (6.17), $H(0)$ is always real (as we already know from our study of the DFT in the previous chapter). This term is, of course, the zero-frequency gain (or dc level). Equations (6.19) indicate that increasing $H(0)$ by some amount has the effect of increasing each $h(n)$ value by a proportionate amount. Further, provided we can make $H(N/2)$ real, we see that $h(n)$ is real regardless of whether N is even or odd, given that we satisfy the complex-conjugate condition.

From equations (6.16) and (6.17):

$$H(m) = \pm|H(m)|e^{-j\pi m(N-1)/N}$$

From the complex-conjugate condition, equation (6.18):

$$|H(m)|e^{\pm j\pi m(N-1)/N} = |H(N-m)|e^{\mp j\pi(N-m)(N-1)/N}$$

This simplifies directly to:

$$|H(m)| = |H(N-m)|e^{\pm j\pi(N-1)}$$

Thus:

$$|H(m)| = \begin{cases} -|H(N-m)| & N \text{ even} \\ \\ |H(N-m)| & N \text{ odd} \end{cases} \tag{6.20}$$

Putting all this together, we have that $H(m)$ must be chosen such that:

- its magnitude corresponds to samples of the desired magnitude response up to the Nyquist frequency;
- it conforms with equation (6.20) above the Nyquist frequency;
- it has phase satisfying equations (6.16) and (6.17).

The resulting filter will then have a finite impulse response and the desired linear-phase property.

Example 2: The low-pass FIR filter design previously performed using FS truncation is to be repeated using the frequency sampling method, again with $N = 17$ coefficients. Find h_5 and show that this (as a consequence of the symmetry which yields linear phase) is equal to h_{11}.

As $(N-1)$ corresponds to the fundamental DFT frequency of $\nu = 2\pi$, index $(N-1)/10$ corresponds to the $\pi/5$ break point. Now, $(N-1)/10 = 1.6$, so that the break frequency falls between frequency samples 1 and 2. Hence, $|H(1)|$ will be 1. By the symmetry condition of equation (6.20) for N odd, $|H(16)|$ will be 1 also. Since 1.6 is non-integer, we can either make $|H(2)|$ – and, by symmetry, $|H(15)|$ also – zero, or interpolate to a value of 0.4. We will do the latter. $|H(3)|$ through $|H(14)|$ are all zero.

From equations (6.16) and (6.17):

$$H(m) = \begin{cases} e^{-16m\pi/17} & m = 1, 16 \\ 0.4e^{-16m\pi/17} & m = 2, 15 \\ 0 & m = 3..14 \end{cases}$$

Substituting into the design equations (6.19) for $n = 5$ and odd N, equal to 17:

$$h_5 = \frac{1}{17}\left\{H(0) + 2Re\left[H(1)e^{10\pi j/17}\right] + 2Re\left[H(2)e^{20\pi j/17}\right]\right\}$$

Since the required zero-frequency gain is 1 (figure 6.4), and we have that $H(0)$ is guaranteed real, put $H(0)$ equal to 1, giving:

$$\begin{aligned} h_5 &= \frac{1}{17}\left\{1+2Re\left[e^{-16\pi j/17}e^{10\pi j/17}\right]+2Re\left[0.4e^{-32\pi j/17}e^{20\pi j/17}\right]\right\} \\ &= \frac{1}{17}\left\{1+2Re\left[e^{-6\pi j/17}\right]+2Re\left[0.4e^{-12\pi j/17}\right]\right\} \\ &= 0.0829 \end{aligned}$$

Because of the linear phase property, $h(n) = h(N-1-n)$. Hence, h_5 should be equal to h_{11}. Let us check that this is so. Using equation (6.19):

$$\begin{aligned} h_{11} &= \frac{1}{17}\left\{1+2Re\left[e^{-16\pi j/17}e^{22\pi j/17}\right]+2Re\left[0.4e^{-32\pi j/17}e^{44\pi j/17}\right]\right\} \\ &= \frac{1}{17}\left\{1+2Re\left[e^{6\pi j/17}\right]+2Re\left[0.4e^{12\pi j/17}\right]\right\} \\ &= 0.0829 \end{aligned}$$

as required. The filter coefficients b_r are, of course, identical to the $h(n)$ values.

Since we are sampling the frequency response in this design method, the question of proper sampling (such as to satisfy the Nyquist criterion) arises. By the principle of time-frequency duality, a sampled function of frequency $H(m)$ can adequately describe the frequency response $H(e^{j\omega T})$ provided the latter is *time*-limited. This is guaranteed for a finite impulse response system. Further, as we found in the previous chapter, the DFT uses the precise, minimum number of sample values to satisfy the Nyquist criterion. Hence, proper sampling is assured in the sense that the frequency response could, in principle, be recovered by ideal interpolation. This, however, is not the same thing as the frequency-sampling filter having the ideal frequency response. As with the Fourier series truncation method, the approximation to the ideal response is improved by increasing the value of N. In practice (as in example 1 above), limits on the value of N may be set by the maximum delay which can be tolerated.

6.5 COMPUTER-BASED DESIGNS

The FS truncation (or 'window') and frequency sampling methods give us very useful ways of designing linear-phase FIR filters. They do, however, suffer (as we have seen) from the problem of rather poor control of certain critical parameters such as the transition width between passband and stopband, and the maximum error between desired and obtained frequency responses. In fact, it is possible to show that a rectangular window for FS truncation achieves the best mean-square error for given duration N, yet it may do so at the expense of an unacceptably large maximum error. We may prefer to minimise this maximum error, for instance. This minimisation criterion is called the *Chebyshev norm.*

A number of iterative design procedures exists. In these, certain parameters are fixed and others are manipulated to yield a linear-phase FIR filter satisfying some optimisation criterion – such as the Chebyshev norm. For instance, one of the ways of increasing stopband attenuation is to widen the transition region. This can be done in a computer-aided design procedure by iteratively adjusting frequency sample values – much as we allowed $|H(2)|$ in our example above to take on a value of 0.4, rather than restricting it to be 1 or 0.

Space precludes any more thorough treatment of these iterative, computer-based design methods. We merely note that perhaps the best known is the Parks-McClellan procedure for designing equiripple FIR linear-phase filters.

6.6 SUMMARY

Finite impulse response (FIR) systems have an impulse response whose sequence values are identically equal to the coefficients of the defining difference equation. Such systems have a number of advantages over their infinite impulse response (IIR) counterparts. First, since there is no feedback – and no poles in the system function – stability is unconditionally guaranteed. Another very important property of FIR filters is that they can be designed to have linear phase. In this case, signals falling within the passband will be output with a time delay equal to the slope of the phase response. Thus, the output is a filtered but undistorted version of the input. Linear phase depends upon the impulse response sequence of the FIR system being symmetric. If the length of this sequence is N, such a filter has a delay of $(N-1)/2$ samples. The disadvantage of FIR designs relative to IIR implementations is that a much larger number of coefficients is generally required in order to achieve good roll-off with low ripple.

FIR filters can be designed by expressing the desired (periodic) magnitude response $|H(e^{j\omega T})|$ as a Fourier series which is then truncated to its first N terms. The Fourier coefficients are then used to define the impulse response values, i.e. the difference equation coefficients. In some applications, N will be fixed by the maximum delay which is desired or which can be tolerated. In other cases, N may be set so as to trade complexity of the filter against error in approximating the desired magnitude response. Linear-phase is ensured because the desired magnitude response is an even function of frequency. Hence, the Fourier coefficients satisfy the well-known symmetry criterion, and the impulse response values do likewise.

Truncation of the Fourier series to N values leads to a problem of ripple, i.e. error between the desired and obtained frequency responses. The maximum error can be reduced (while leaving N unchanged) by appropriate choice of a window function to taper the Fourier series values, rather than truncating them abruptly (which corresponds to use of a rectangular window). This reduction in the maximum error is generally at the expense of the width of the transition band (relative to use of a rectangular window).

An alternative to FS truncation samples the desired magnitude response in the baseband, and considers these sample values to be the magnitudes of the DFT of the filter's impulse response. Corresponding phase values are chosen so as to ensure linear phase, and to yield real impulse response values.

The obtained filter responses only approximate the desired magnitude responses. Computer-aided FIR design methods use these techniques as a starting point, examine the goodness of fit to the desired characteristics, and iteratively change filter parameters until an acceptable approximation is found, based on some optimisation criterion.

6.7 EXERCISES

6.1 Show that a constant time (or group) delay function implies a uniform time delay for all sinusoids irrespective of frequency. Why is the negative sign necessary in equation (6.2)?

6.2 Is a linear-phase response ensured if an FIR filter has an *anti*-symmetric impulse response? Will the system function be purely real, purely imaginary or complex in this case?

6.3 What constraints does the linear-phase property place on the positions of zeros on the z-plane?

6.4 Using the Fourier series truncation technique and a rectangular window, design a band-pass filter that approximates the ideal $|H_d(e^{j\omega T})| = 1$ over the frequency range $100 < f < 250$ Hz and $|H_d(e^{j\omega T})| = 0$ elsewhere, with a sampling frequency of 1 kHz. Limit the duration of the impulse response to 40 ms. Using a spreadsheet or otherwise, evaluate the obtained magnitude response and compare it with the desired one.

6.5 Complete the design of the low-pass filter in example 2 above, i.e. find the $h(n)$ values in addition to h_5 and h_{11}. Using a spreadsheet or otherwise, evaluate the magnitude response and compare it to that for example 1 (c.f. figure 6.5).

6.6 Repeat the design of the low-pass filter in example 2 above using $N = 18$ coefficients rather than 17.

6.7 Repeat the design of exercise 6.4 using a Hamming window.

6.8 Plot the Bartlett window, $w(n)$, versus n. Find its spectrum and compare it with that for the rectangular and Hamming windows (exercise 5.12).

7

The fast Fourier transform

In chapter 5, we studied the discrete Fourier transform in some depth, concentrating on simple, *deterministic* DT signals – those which can be described by a relatively simple mathematical formula – of short length, N, so that the DFT could be computed without undue difficulty. The transform allowed us to represent a signal in terms of its frequency spectrum. In many practical cases, the frequency-domain representation is of considerable interest for its own sake, in giving a different view of the signal from the time-domain waveform that makes different features explicit. For instance, the resonant frequencies of a speech signal or of some vibration of a mechanical structure may not be especially easy to see in the time-domain waveform, yet will appear as distinct peaks in the DFT magnitude (or power) spectrum. In practice, sampled-data signals such as these will usually be extensive in time. Hence, since many data points must be processed, we require efficient computational procedures for evaluating the DFT.

In many situations, the DFT spectrum is used only as an intermediate representation in a process such as interpolation, linear filtering, FIR filter design or system identification. Most often, we are essentially replacing time-domain convolution by the simpler operation of multiplication in the frequency domain. In order to do this, however, we need to perform both a forward and an inverse time-to-frequency transform; hence, any overall computational saving is critically dependent upon the efficiency of the DFT and IDFT algorithms. Furthermore, many signals of interest (such as speech) change more or less slowly over time, and these changes may need to characterised. Hence, a long sequence may be processed as a series of smaller (length-N) blocks (section 5.7) to give a time-frequency representation. In this case, the DFT may be invoked many times over

– once per block. The same block-based processing will be used for the linear filtering of a long sequence, with both the DFT and IDFT used many times over. Recall, however, that the forward and inverse transforms (equations (5.4) and (5.8)) have remarkably similar form so that essentially the same algorithm can be used for both.

For all the above reasons, the efficiency of DFT algorithms has been a focus of considerable interest in the past. Fortunately, the symmetry (complex-conjugate) and periodicity properties of the DFT that we explored in some detail in chapter 5 can be exploited to good effect. As we shall see, these properties allow us to divide the computation of a length-N DFT down into the computation of two DFTs of smaller length with consequent savings – a process called *decimation*. The strategy can be repeated iteratively to produce further computational savings which increase dramatically as N increases. Such *divide-and-conquer* strategies are very powerful and, in various guises, are commonly employed in efficient algorithmic implementations of all sorts.

There are very many different, detailed ways that the DFT properties can be exploited to achieve efficiencies and so produce a *fast* Fourier transform (FFT) algorithm. Hence, there is a whole class of FFT algorithms for computing the DFT but it is usual to refer to them all collectively as *the* fast Fourier transform. The variant that we will concentrate on relies on N being a power of 2, so that the decimation divides the sequence (or sub-sequences) into two. This is the class of *radix-2* algorithms. Two other, basic classes of FFT can be identified: those in which $x(n)$ is divided into smaller sequences and those in which $X(m)$ is so divided. These are referred to as *decimation-in-time* and *decimation-in-frequency* respectively.

From chapter 5, equation (5.4), the DFT of a length-N sequence $x(n)$ is defined as:

$$X(m) = \sum_{n=0}^{N-1} x(n) e^{-2\pi jmn/N} \qquad 0 \le m < (N-1)$$

We previously assumed that $x(n)$ – as a sampled-data version of a CT signal $x(t)$ – would usually be real. If, however, we are to use essentially the same algorithm for the forward and inverse transforms, and given that $X(m)$ is generally complex, it makes sense to treat $x(n)$ as complex. That is, we develop the FFT on the assumption that the input sequence is complex. Note carefully that the symmetry and periodicity properties developed in chapter 5 were for $x(n)$ real. Hence, the properties we use here will be similar but not identical to those in chapter 5 (see exercise 7.1).

7.1 Direct evaluation of the DFT

For each value of m, with $x(n)$ complex, the direct evaluation of $X(m)$ using equation (5.4) involves N complex multiplications and $(N-1)$ complex additions. In terms of corresponding real operations, there are $4N$ real multiplications and $(4N - 1)$ real additions. In general, there are 3 additions to evaluate each complex multiplication plus 1 addition per multiplication, giving $4(N-1)$ additions, except that there are also 3 additions required for the Nth value, giving $4(N - 1) + 3 = (4N - 1)$ additions in all. This must be repeated for all m values, of which there are N in total. Since addition is a considerably quicker and easier computational operation than multiplication, the efficiency of the algorithm is dominated by the number of multiplications. There are N^2 complex multiplications, or $4N^2$ real multiplications. We say that the *computational complexity* of the DFT when calculated directly is *of order* N^2, denoted $O(N^2)$. This notation – like that introduced in chapter 2 for the approximation error of a discrete-time differentiator or integrator – indicates that the efficiency of the algorithm depends primarily on some number of basic operations proportional to N^2 but does not specify a constant of proportionality. The significance of this is that as N increases by a factor k, the required computation goes up by the square of k, irrespective of the value of the constant. Thus, $O()$ is sometimes called the *growth function*.

We will now consider FFT algorithms which reduce the necessary computation by exploiting certain properties of the DFT. When dealing with the FFT, it is usual to replace the (principal) Nth root of unity in equation (5.4) by:

$$W_N = e^{-2\pi j/N} \tag{7.1}$$

to give:

$$X(m) = \sum_{n=0}^{N-1} x(n) W_N^{mn} \tag{7.2}$$

In FFT terminology, equation (7.2) is called the *N-point* transform, and W_N (or, indeed, any of its integer powers, W_N^{mn}) is called a *phase factor*.

The specific properties to be used in finding efficient algorithms for the computation of the DFT are the symmetry (complex-conjugate) and periodicity properties.

7.2 RADIX-2 ALGORITHMS

Here, we consider the case where $N = 2^\nu$ (ν is a positive integer).

7.2.1 Decimation-in-time

Since $N = 2^\nu$ is even, we can split the computation of the N-point transform in equation (7.2) into the evaluation of two *interleaved* $N/2$-point sub-sequences:

$$X(m) = \sum_{n_e=0}^{N-2} x(n_e) W_N^{mn_e} + \sum_{n_o=1}^{N-1} x(n_o) W_N^{mn_o}$$

Here, the first summation is indexed by the even values, $n_e = 0, 2, \ldots, (N-2)$, of n while the second is indexed by the odd values, $n_o = 1, 3, \ldots, (N-1)$.

Substituting $n_e = 2r$ and $n_o = (2r+1)$:

$$\begin{aligned} X(m) &= \sum_{r=0}^{N/2-1} x(2r) W_N^{2mr} + \sum_{r=0}^{N/2-1} x(2r+1) W_N^{m(2r+1)} \\ &= \sum_{r=0}^{N/2-1} x(2r) \left(W_N^2\right)^{mr} + W_N^m \sum_{r=0}^{N/2-1} x(2r+1) \left(W_N^2\right)^{mr} \end{aligned}$$

But, from equation (7.1):

$$\begin{aligned} W_N^2 &= e^{-4\pi j/N} = e^{-2\pi j\left(\frac{N}{2}\right)} \\ &= W_{N/2} \end{aligned} \tag{7.3}$$

Hence:

$$\begin{aligned} X(m) &= \sum_{r=0}^{N/2-1} x(2r) W_{N/2}^{mr} + W_N^m \sum_{r=0}^{N/2-1} x(2r+1) W_{N/2}^{mr} \\ &= F_e(m) + W_N^m F_o(m) \qquad 0 \le m < (N-1) \end{aligned} \tag{7.4}$$

where $F_e(m)$ and $F_o(m)$ are the $N/2$-point DFTs of the even- and odd-indexed values of $x(n)$ respectively.

In equation (7.4), $X(m)$ is evaluated over the range $0 \le m < (N-1)$. However, from the periodicity property with period $N/2$:

$$\begin{aligned} F_e(m+N/2) &= F_e(m) \\ \text{and} \quad F_o(m+N/2) &= F_o(m) \end{aligned} \tag{7.5}$$

Hence, we only need evaluate the two summations in equation (7.4) over the range $0 \le m < (N/2-1)$ in order to find values for $X(m)$ over the entire range.

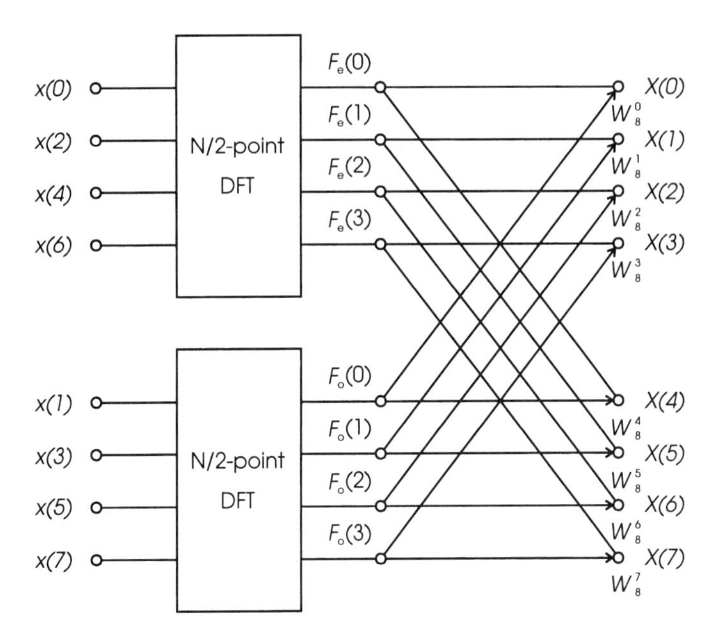

Fig. 7.1 Signal flow graph for the first stage of the decimation-in-time FFT computations with $N = 8$. The input sequence $x(n)$ is split into two (even- and odd-indexed) sub-sequences and an $N/2$-point DFT performed on each. Signals on lines shown arrowed are multiplied by the phase factor W_N^m as indicated. Transmission along lines without arrows is unity-weighted.

The computations in equation (7.4) can be visualised as in the *signal flow* graph of figure 7.1 for the case of $N = 8$. Such graphs are an alternative to the block diagrams introduced in chapter 2 as a means of visualising the computations carried out by a DT system. In the graph, signals on lines into a node are summed to produce the signal flowing out of the node. Signal lines with arrows indicate that the variable is weighted by multiplication by the phase factor W_N^m as indicated, while lines without arrows denote that the signal is unweighted. It is perhaps worth noting at this point that W_N^0 is identically equal to 1, so that the multiplication in this case is redundant.

We previously found that N^2 complex multiplications were required to compute the DFT directly. With the computation as shown in figure 7.1, we need to evaluate two $N/2$-point sub-sequences. Hence, the required number of complex multiplications is $2(N/2)^2$, or $(N^2/2)$. Also, N multiplications by the phase factors W_N^m are necessary, giving $(N^2/2) + N$ multiplications in total. For the 8-point transform considered here, 32 + 8 = 40 complex multiplications are needed, compared to 64 for direct computation

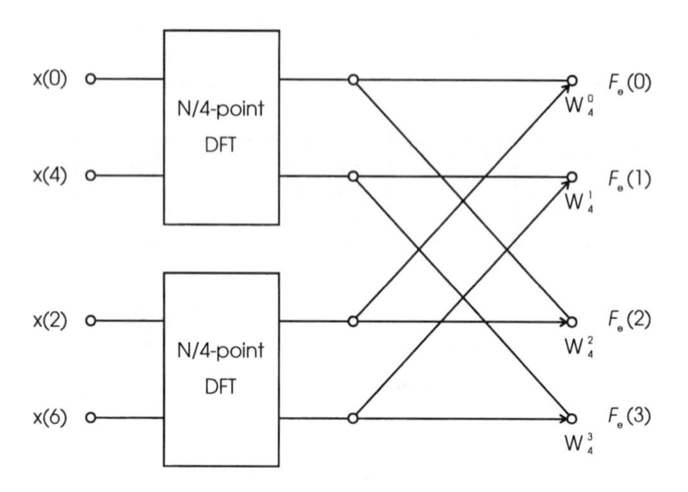

Fig. 7.2 Signal flow graph for the $N/2 = 4$ even-indexed sub-sequence computations in the decimation-in-time FFT with $N = 8$.

of the DFT. When N is large, it can be ignored in comparison with $N^2/2$ and the amount of computation is approximately halved relative to direct evaluation.

Because N is an integer power (ν) of 2, however, the sub-sequences $x_e(n)$ and $x_o(n)$ are also of even length ($N/2$). Hence, the decimation process described above can be applied again to each sub-sequence. Figure 7.2 depicts this for the even-indexed sub-sequence $F_e(m)$ $(0 \leq m < (N/2 - 1))$ of equation (7.4). Similar computations are required for $F_o(m)$, of course. Each of these involves the evaluation of two $N/4$-point DFTs and, again, $N/4$ is even. In principle, then, the decimation step can be applied once more. However, in this case ($N = 8$), the length of the sub-sequence is $N/4 = 2$ and the 2-point transform can be found very simply, since the summations in equation (7.4) then trivially reduce to single values. Figure 7.3 shows the computations required to evaluate the $N/4$-point DFT of $x(0)$ and $x(4)$ as in figure 7.2. Note that it is immaterial whether we use the phase factors for the reduced (2-point) or full (8-point) transform, because their periodicity means that they are identical. Also, they are equal to ± 1 so that no complex multiplications whatsoever are required.

For some value of N which is an integer power (ν) of 2, we can apply $\nu = \log_2 N$ stages of decimation: the last stage will correspond to evaluation of the 2-point transform. Now, in figure 7.1 there are $N = 8$ complex multiplications by the phase factors for the 8-point transform. In figure 7.2, however, there are $N/2 = 4$ such operations for the even-indexed inputs but another 4 complex multiplications are needed for the odd-

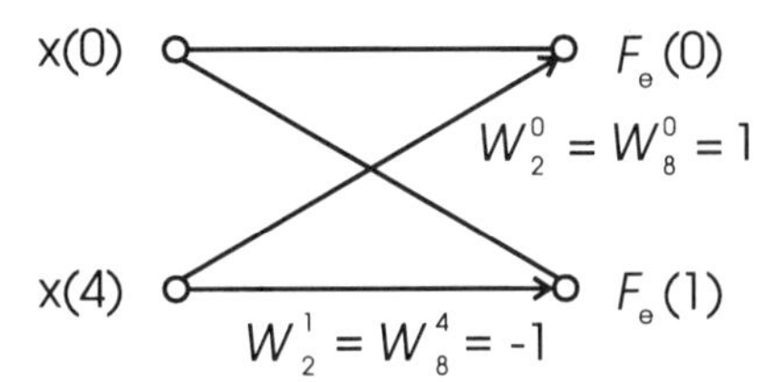

Fig. 7.3 Signal flow graph for the $N/4 = 2$ even-indexed sub-sequence computations in the decimation-in-time FFT with $N = 8$.

indexed inputs. Finally, in figure 7.3, we require 2 multiplications for inputs $x(0)$ and $x(4)$ or, again, 8 in total. However, these multiplications are real – either by 1 or -1. Hence, for general N, we need N complex multiplications for each stage except the last. Ignoring the exceptional nature of the last stage, we require something like $N \log_2 N$ complex multiplications. This represents a very significant saving of computation relative to direct evaluation of the DFT for even quite moderate values of N. However, yet a further saving is possible.

We will denote the array of values resulting from the ith stage of computation as $A_i(l)$, $0 \leq i < \nu$ and $0 \leq l < (N-1)$. Here, $A_0(l)$ represents the array of input values. Now, because of the way that we successively split the input array, it turns out to be computationally advantageous to re-order the input data for our example 8-point transform as follows:

$$\begin{array}{llllllll} A_0(0) = x(0) & A_0(2) = x(2) & A_0(4) = x(1) & A_0(6) = x(3) \\ A_0(1) = x(4) & A_0(3) = x(6) & A_0(5) = x(5) & A_0(7) = x(7) \end{array}$$

The reason for this ordering can be understood from figure 7.3, where we see that the 2-point DFT of $x(0)$ and $x(4)$ is to be computed as part of the first stage of the algorithm. Similarly, the 2-point DFT of $x(2)$ and $x(6)$ is required, and so on. This re-sequencing of the data is known as *bit-reversal*, since the index of each $x()$ is obtained by reversing the ν-bit ($\nu = 3$) binary representation of the corresponding $A_0()$ index. For instance, $6_{10} = 110_2$; reversing the binary representation gives $011_2 = 3_{10}$, so that $A_0(3) = x(6)$ and $A_0(6) = x(3)$.

With this re-sequencing, the basic FFT operation is as depicted in figure 7.4 which can be seen as a more general version of figure 7.3. Because of its graphical appearance, this is called the FFT *butterfly*. The necessary computations are of the form:

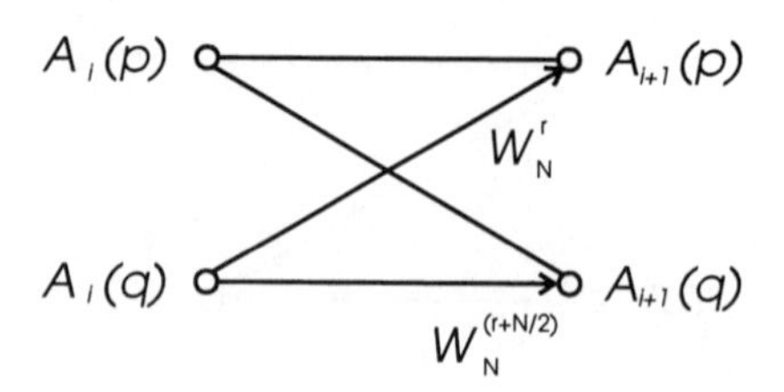

Fig. 7.4 Signal flow graph for the FFT butterfly computation.

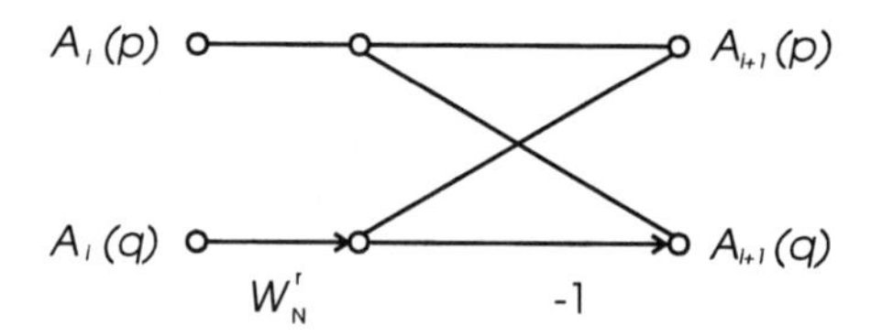

Fig. 7.5 The two complex multiplications in the FFT butterfly can be replaced by one complex multiplication and a sign reversal.

$$\left.\begin{aligned} A_{i+1}(p) &= A_i(p) + W_N^r A_i(q) \\ A_{i+1}(q) &= A_i(p) + W_N^{r+N/2} A_i(q) \end{aligned}\right\} \quad 0 \leq m < (N/2 - 1) \tag{7.6}$$

where the index r depends upon the exact position of the butterfly in the computational structure described by the signal flow graph. However:

$$W_N^{r+N/2} = e^{-2\pi jr/N} e^{-\pi j} = -W_N^r \tag{7.7}$$

Hence, equations (7.6) become:

$$\begin{aligned} A_{i+1}(p) &= A_i(p) + W_N^r A_i(q) \\ A_{i+1}(q) &= A_i(p) - W_N^r A_i(q) \end{aligned}$$

so that only one complex multiplication is required per butterfly, as depicted in figure 7.5, thus halving the total number of such operations relative to our original algorithm. As there are $N/2$ butterflies at each stage, and $\log_2 N$ stages overall, the number of complex multiplications is $(N/2)\log_2 N$. Note that the growth function of both variants of the FFT is the same, $O(N \log_2 N)$, but the constant of proportionality is lower here, being half that of the earlier algorithm.

In the form that we have described it here, the DFT takes N complex input values and transforms them to N complex output values. From figure 7.5, we see that the bit-reversal re-ordering of the input means that butterfly input and output values for the same index (p or q) appear on the same horizontal row. As a consequence, we need only one (complex) length-N array to perform the complete transformation. Once $A_{i+1}(p)$ and $A_{i+1}(q)$ are computed, they can replace $A_i(p)$ and $A_i(q)$ (which are no longer needed) in the complex array. This is referred to as *in-place computation*. It can make a significant impact on efficiency, since it becomes feasible to store all the necessary data on the processor chip avoiding any need for external memory operations. However, because of the bit-reversal, the index values are accessed in non-sequential order. Hence, the on-chip memory needs to be random-access. Modern digital-signal processing (DSP) chips have an architecture purpose-designed for just this sort of computation, although the amount of on-chip memory is limited by cost.

The decimation-in-time algorithm with bit-reversal and in-place computation that we have described is the 'classical' FFT algorithm. It is essentially the variant introduced by Cooley and Tukey in 1965, the availability of which revolutionised digital signal processing. A minor difference, however, is that the Cooley-Tukey algorithm preserved the input sequence in normal order but produced a bit-reversed output.

7.2.2 Decimation-in-frequency

The decimation-in-time algorithm just described iteratively divides the input sequence into ever smaller sub-sequences until we arrive at a basic 'unit' – the FFT butterfly – for which the necessary computation is very simple. As an alternative, it is possible to divide the output sequence in a similar fashion to give a decimation-in-frequency algorithm.

The first step is to split the input sequence into two non-interleaved halves as follows:

$$X(m) = \sum_{n=0}^{N/2-1} x(n) W_N^{mn} + \sum_{n=N/2}^{N-1} x(n) W_N^{mn}$$

Replacing n in the second summation by $(n + N/2)$:

$$X(m) \quad = \quad \sum_{n=0}^{N/2-1} x(n) W_N^{mn} + \sum_{n=0}^{N/2-1} x\left(n + \frac{N}{2}\right) W_N^{m(n+N/2)}$$

$$= \sum_{n=0}^{N/2-1} x(n) W_N^{mn} + W_N^{m(N/2)} \sum_{n=0}^{N/2-1} x\left(n + \frac{N}{2}\right) W_N^{mn}$$

But, $W_N^{N/2} = -1$ so that:

$$X(m) = \sum_{n=0}^{N/2-1} \left[x(n) + (-1)^m x\left(n + \frac{N}{2}\right)\right] W_N^{mn}$$

We can now divide the output sequence $X(m)$ into two interleaved parts, one with m even and the other with m odd:

$$X(2r) = \sum_{n=0}^{N/2-1} \left[x(n) + x\left(n + \frac{N}{2}\right)\right] W_N^{2rn}$$

$$X(2r+1) = \sum_{n=0}^{N/2-1} \left[x(n) - x\left(n + \frac{N}{2}\right)\right] W_N^{n} W_N^{2rn}$$

Since $W_N^{2rn} = W_{N/2}^{rn}$, each of these has the form of an $N/2$-point DFT. In the case of $X(2r)$ (m even), the DFT is of the sum of corresponding values (i.e. spaced by $N/2$) of the first- and second-half sub-sequences. In the case of $X(2r+1)$ (m odd), the DFT is of the difference of corresponding values of the first- and second-half sub-sequences multiplied by a phase factor.

The signal flow graph for the case of $N = 8$ is shown in figure 7.6. Clearly, we can apply the same division strategy to the two $N/2$-point DFTs, much as we did in the decimation-in-time algorithm. The detailed working is left to the reader but, ultimately, we arrive at the butterfly computations of figure 7.7. Again, the number of complex multiplications (with in-place computation) is $(N/2)\log_2 N$.

7.2.3 Time-space trades

There is some scope for trading computation time against storage space requirements in FFT algorithms. We consider here a couple of instances of this.

If the computations are in-place, then we need only $2N$ storage locations for the (single) array of N complex numbers. At each stage of computation, the bit-reversal enables us to overwrite an input value with its corresponding output value once the latter is obtained. If, however, we use two such arrays, i.e. $4N$ storage locations, then indexing of the array(s) can be simplified. At one particular stage of the FFT, then, we will read from one array containing the input data to that stage and write the computed

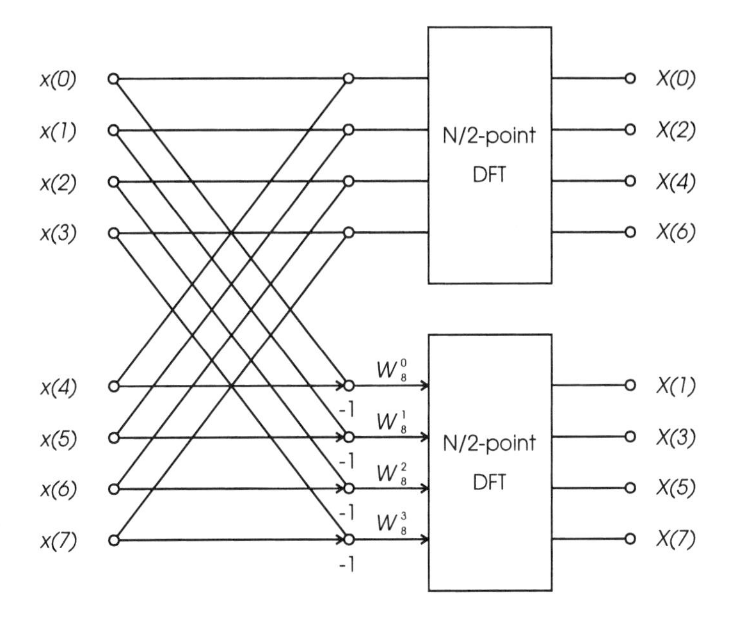

Fig. 7.6 Signal flow graph for the first stage of the decimation-in-frequency FFT computations with $N = 8$.

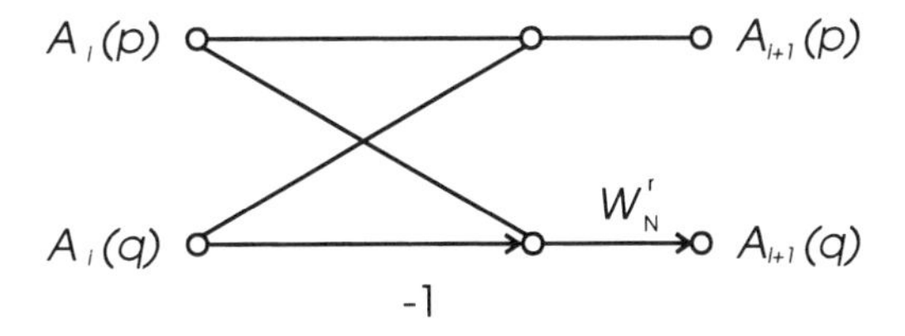

Fig. 7.7 Butterfly computations for the decimation-in-frequency FFT algorithm.

output data to the other array. At the next stage, we reverse the rôle of the two arrays. Since there is never any danger of overwriting values that will be needed later, there is no necessity for bit-reversal.

The N-point transform requires phase factors $W_N^m, 0 \leq m < (N/2 - 1)$, to be available. These can either be pre-computed and stored in a look-up table, or evaluated as they are required. The former has the advantage of speed but requires $N/2$ complex storage locations (or N real locations) to be available. Symmetry properties can be used to reduce this number but there is a consequential increase in the cost of computing the appropriate index. The phase factors may be required in either normal or bit-reversed order. In the latter case, they can either be stored in this order, or stored in normal order but retrieved in bit-reversed order implying more complex indexing.

7.2.4 Inverse FFT

As emphasised earlier, the inverse DFT has remarkably similar form to the forward DFT. It can be written:

$$x(n) = \frac{1}{N} \sum_{m=0}^{N-1} X(m) W_N^{-mn}$$

The only differences from the DFT are the scaling factor $1/N$ and change of sign of the phase factors – and, of course, the input and output sequences are swapped. Hence, essentially the same FFT algorithm can be used for both the DFT and the IDFT.

If we take the signal flow graph for the decimation-in-time algorithm, reverse the direction of signal flow so that the input and output sequences are interchanged, change the sign of the phase factors, and divide the output by $1/N$, we obtain a fast, decimation-in-frequency structure for computing the IDFT (exercise 7.7). Similarly, we can convert the decimation-in-frequency algorithm for the forward DFT into a decimation-in-time structure for the IDFT. By these means, FFT algorithms for the IDFT are easily found. Given an FFT algorithm for complex input $x(n)$, however, it is possible to use the same code for both forward and inverse transforms. Since $x(n)$ is complex, we can write:

$$\begin{aligned} x(n) &= Re\,[x(n)] + j\,Im\,[x(n)] \\ &= x_R(n) + jx_I(n) \end{aligned}$$

So the complex conjugate of $x(n)$ is:

$$\begin{aligned} \tilde{x}(n) &= x_R(n) - jx_I(n) \\ \text{and} \quad j\tilde{x}(n) &= x_I(n) + jx_R(n) \end{aligned}$$

That is, taking the complex conjugate and multiplying by j corresponds to swapping the real and imaginary parts. But, from the IDFT:

$$\begin{aligned} x(n) &= \frac{1}{N}\sum_{m=0}^{N-1} X(m) W_N^{-mn} \\ \text{so that} \quad j\tilde{x}(n) &= \frac{1}{N}\sum_{m=0}^{N-1} j\tilde{X}(m) W_N^{mn} \\ &= \frac{1}{N}\mathcal{D}\left[j\tilde{X}(m)\right] \end{aligned}$$

Hence $x(n)$ – the IDFT of $X(m)$ – can be found by interchanging real and imaginary parts of $X(m)$, taking the forward transform using the FFT, scaling the output by $1/N$ and interchanging the real and imaginary parts of the output. The code for the forward transform is used unaltered.

7.3 ALGORITHMS FOR COMPOSITE-N

The radix-2 FFT algorithms achieve their computational efficiency by iteratively dividing the N-point sequence in two. As such, they rely on N being an integer power (ν) of 2. When processing a long sequence block-by-block, there is not usually any difficulty in choosing the window length to be an integer power of 2. Sometimes, however, this may prove problematic – for instance, if we wish to process an integer number of periods of a periodic (or quasi-periodic) signal, as in pitch-synchronous speech processing. In this case, if the window length is not an integer power of 2, the sequence can simply be padded with the requisite number of zeros such that N becomes equal to 2^ν. As we found in section 5.8, the only effect will be to increase the resolution of the DFT spectrum by reducing the frequency spacing of the sample values.

Even so, there may still be occasions where $N \neq 2^\nu$ for some reason or another, and it is not possible or convenient to make it so. As these occasions are rare, we will not consider them in any depth here – beyond remarking that the divide-and-conquer, or decimation, strategy can still be usefully applied when N is highly composite (i.e. has many prime factors) and justifying this remark.

As an illustration, suppose that $N = 9 = 3\times3$. We can then divide the input sequence $x(n)$ into 3 interleaved sequences, each of length 3:

$$\begin{aligned} x_1(n) &= \{x(0), x(3), x(6)\} \\ x_2(n) &= \{x(1), x(4), x(7)\} \\ x_3(n) &= \{x(2), x(5), x(8)\} \end{aligned}$$

so that we have to evaluate and combine three 3-point DFTs. Each of these will require $3^2 = 9$ complex multiplications, or 27 for all 3 DFTs. In addition, another 9 complex multiplications are required for the combination, giving a total of 36 in all. This is a considerable saving relative to $N^2 = 81$.

Clearly, a general composite-N algorithm will be complicated by the necessity to decide on the factors to be used in the decomposition. In principle, the decomposition should be into factors which are prime for maximal efficiency.

7.4 THE DFT OF REAL DATA

We have developed the FFT in this chapter on the assumption that the input data are complex. That is, the input will consist of an array of N real values and another array of N imaginary values. The main advantage of so doing – apart from increased generality – is that we can then exploit the essential similarity of the forward and inverse DFTs, using the same (forward transform) code as the basis of both. In the vast majority of practical applications, however, the input data are real. In this case, the use of a general, complex-data FFT algorithm is inherently inefficient: the array of imaginary input values will be identically zero yet the algorithm will treat them exactly as if they were non-zero.

Recognising that the imaginary input values are identically zero for a real sequence, we can very easily implement the FFT to accept a length-N array of real values only. The efficiency savings in so doing are minimal, however, since it is only at the very first, input stage that the data values are real. At all subsequent stages of computation, provision must be made to handle complex data. At the output, we will have a length-N sequence of ($2N$) complex values. If the input data are real, however, the real and imaginary parts of the output sequence will have even and odd symmetry respectively about the folding frequency, as found in chapter 5, so that half of the output values can be trivially inferred from the other half. For this reason, it is common to arrange for the output to contain just the $N/2$ real values and the $N/2$ imaginary values between zero frequency and the folding frequency, $f_s/2$. Given that half of the $2N$ complex output values need not be computed, there is obvious scope for further efficiencies to be made (exercise 7.8). The reader is warned that a variety of FFT algorithms is available in software packages and published literature, and the authors of these algorithms take different approaches to the representation of the input and output sequences when the input is real. In order to use these properly, it is essential to understand precisely the representation(s) being employed.

Given a general N-point algorithm intended to transform a length-N complex data sequence, however, it is possible to use it more efficiently to transform a length-$2N$ real data sequence. As in the derivation of the decimation-in-time algorithm, we can split the input sequence $x(n)$ into interleaved even- and odd-indexed values, $u(n)$ and $v(n)$, such that:

$$\begin{aligned} u(n) &= x(2n) \\ v(n) &= x(2n+1) \end{aligned}$$

each of which is a length-N sequence if $x(n)$ is length-$2N$. We can then form a new length-N complex sequence as:

$$g(n) = u(n) + jv(n)$$

We can use the general FFT for complex input to take the DFT of this sequence to give $G(m)$, from which we wish to find $X(m)$. Now:

$$\begin{aligned} u(n) &= \frac{g(n) + \tilde{g}(n)}{2} \\ v(n) &= \frac{g(n) - \tilde{g}(n)}{2j} \end{aligned}$$

so that:

$$\begin{aligned} U(m) &= \frac{1}{2}\left(\mathcal{D}[g(n)] + \mathcal{D}[\tilde{g}(n)]\right) \\ V(m) &= \frac{1}{2j}\left(\mathcal{D}[g(n)] - \mathcal{D}[\tilde{g}(n)]\right) \end{aligned}$$

In order to find the DFT of $\tilde{g}(n)$, we note that:

$$\begin{aligned} G(m) &= \mathcal{D}[g(n)] \\ &= \sum_{n=0}^{N-1} g(n) e^{-2\pi jmn/N} \end{aligned}$$

so that by symmetry

$$\begin{aligned} G(N-m) &= \sum_{n=0}^{N-1} g(n) e^{-2\pi j(N-m)n/N} \\ &= \sum_{n=0}^{N-1} g(n) e^{2\pi jmn/N} \end{aligned}$$

and

$$\begin{aligned} \tilde{G}(N-m) &= \sum_{n=0}^{N-1} \tilde{g}(n) e^{-2\pi jmn/N} \\ &= \mathcal{D}[\tilde{g}(n)] \end{aligned}$$

Hence:

$$\begin{aligned} U(m) &= \frac{1}{2}\left[G(m) + \tilde{G}(N-m)\right] \\ V(m) &= \frac{1}{2j}\left[G(m) - \tilde{G}(N-m)\right] \end{aligned}$$

requiring N real multiplications to find $U(m)$ and N complex multiplications to find $V(m)$. Having obtained $U(m)$ and $V(m)$ from $G(m)$ using these equations, we can now find $X(m)$ in terms of $U(m)$ and $V(m)$ by splitting the length-$2N$ sequence into even- and odd-indexed (length-N) sequences just as we did for the decimation-in-time FFT algorithm:

$$\begin{aligned} X(m) &= \sum_{n=0}^{N-1} x(2n) W_{2N}^{m(2n)} + \sum_{n=0}^{N-1} x(2n+1) W_{2N}^{m(2n+1)} \\ &= \sum_{n=0}^{N-1} u(n) W_{2N}^{2mn} + W_{2n}^{m} \sum_{n=0}^{N-1} v(n) W_{2N}^{2mn} \\ &= \sum_{n=0}^{N-1} u(n) W_{N}^{mn} + W_{2N}^{m} \sum_{n=0}^{N-1} v(n) W_{N}^{mn} \end{aligned}$$

(since $W_{2N}^2 = W_N$)

Hence:

$$\left.\begin{aligned} X(m) &= U(m) + W_{2N}^{m} V(m) \\ X(m+N) &= U(m) - W_{2N}^{m} V(m) \end{aligned}\right\} \quad 0 \leq m < (N-1)$$

where we have again used the periodicity and symmetry properties:

$$W_{2N}^{m+N} = e^{-2\pi jm/2N} e^{-\pi j} = -W_{2N}^{m}$$

Only N complex multiplications are required to evaluate the $W_{2N}^{m} V(m)$ terms. This means that a general (complex-input) N-point FFT can be used to find the DFT of a length-$2N$ real sequence with a modest amount ($2N$ complex multiplications) of additional computation.

7.5 SUMMARY

The discrete Fourier transform (DFT) occupies a centrally important place in digital signal processing. Not only is the frequency-domain representation (in terms of the DFT spectrum) of a signal of interest in its own right, the spectrum is used as an intermediate representation in linear filtering, FIR filter design, system identification etc. Hence, the efficient computation of the DFT is a matter of considerable practical interest. The DFT can be evaluated by working directly from the defining summation. This involves N^2 complex multiplications and $N(N-1)$ complex additions, with the number of multiplications dominating the processing time (at least on a general-purpose computer). The computational complexity of the direct evaluation is, therefore, of order $O(N^2)$. Special-purpose digital-signal processing (DSP) architectures – optimised for operations like complex multiplication – can be employed for real-time DFT applications, but the number of complex multiplications remains a useful indicator of complexity. Fortunately, there exists a class of efficient algorithms for computing the DFT known collectively (if slightly inaccurately) as the fast Fourier transform (FFT) which reduces complexity significantly below $O(N^2)$.

The complex multiplications in DFT computations involve so-called phase factors, which are actually (integer powers of) Nth roots of unity. As such, these factors have important periodicity and symmetry properties, and certain of them are identically equal to ± 1. FFT algorithms effect savings by recognising that only the phase factors corresponding to the principal Nth root need be evaluated. This allows a classical divide-and-conquer strategy to the reduction of computational complexity to be employed. Either the input sequence or the output sequence is divided into sub-sequences which are processed separately. This strategy exploits the inequality $r^p + r^q < r^{(p+q)}$ when r, p

and q are positive integers (greater than 1). The main focus of this chapter has been radix-2 ($r = 2$) algorithms for which the input sequence length, N, must be an integer power, ν, of 2. For generality, we have developed the topic assuming the input sequence to be complex (having $2N$ degrees of freedom) while recognising that practical input signals will be real.

In FFT parlance, the divide-and-conquer strategy is known as decimation. Division of the input sequence leads to a so-called decimation-in-time algorithm, while division of the output sequence yields a decimation-in-frequency algorithm: the reason for this terminology is obvious. The division is into interleaved sub-sequences with, in the radix-2 case, one sub-sequence holding the even-indexed values and the other the odd-indexed values. For the radix-2 algorithm, the decimation step can be applied iteratively, producing two sub-sequences which are half the length of their parent (sub-)sequence, until we arrive at length-2 sub-sequences, whose DFTs can be evaluated relatively trivially. The structure of these computations can be visualised using a signal flow graph, similar to the block diagram used in chapter 2 to describe the operation of DT systems, which is a valuable aid to the design of FFT algorithms. Because of the appearance of its signal flow graph, the length-2 DFT computation – which involves just two complex multiplications – is called the FFT butterfly. There are $N/2$ butterflies at each stage (iteration) of decimation, and $\log_2 N$ stages, so that there are $N \log_{2_N}$ complex multiplications in total. For reasonably large N, this number is much less than N^2. However, we can again employ the periodicity/symmetry properties of the phase factors to reduce the number of complex multiplications per butterfly to one – plus a sign reversal. Hence, the number of complex multiplications halves to become $(N/2) \log_2 N$.

Inspection of the signal flow graph for the decimation-in-time algorithm suggests that space savings can be made by re-ordering the input sequence according to so-called bit-reversal of the index values. This allows intermediate values computed as outputs from a given stage to overwrite the inputs to the stage, which are no longer needed. Hence, only a single (length-$2N$) array is required. This is referred to as in-place computation. Such space savings assume major importance when FFT processing is implemented on a DSP chip whose fast, on-chip memory is limited by cost. These savings are at the expense of more complex array addressing, implying the existence of a time-space trade. Another instance where such trades can be made is in the pre-computation and storage (in a look-up table) of the phase factors, versus evaluating the factors as and when they are needed.

A useful duality exists between the decimation-in-time form of the FFT and the decimation-in-frequency form, in the following sense. If we take the signal flow graph for the decimation-in-time algorithm, reverse the direction of signal flow so that the input and output sequences are interchanged, change the sign of the phase factors, and divide the output by $1/N$, we obtain a decimation-in-frequency structure for computing the IDFT. The dual set of operations produces a decimation-in-time structure for computing the IDFT from the decimation-in-frequency algorithm for the forward transform. However, one advantage to design the FFT to accept a complex input sequence is that essentially the same code can then be used for both the forward and inverse transforms. This relies on the fact that interchanging the real and imaginary parts of a complex number corresponds to conjugation and multiplication by j. Given code for the forward transform, then, it can be utilised to find the IDFT by interchanging the real and imaginary parts of the $X(m)$ input sequence, scaling the outputs by $1/N$ and interchanging the real and imaginary parts of the output sequence to yield $x(n)$.

The divide-and-conquer strategy works especially well in the case of the radix-2 FFT, where N is an integer power, ν, of 2. Accordingly, it is usual practice to arrange that blocks for FFT processing are of length $N = 2^\nu$, either directly by selecting precisely this amount of data from the input sequence or indirectly by zero-padding up to length 2^ν. When this is not possible, however, the divide-and-conquer strategy can still be applied with advantage if N is highly composite. This was illustrated for $N = 9$, with the input sequence decimated into 3 interleaved sequences, each of length 3. The required number of complex multiplications is thereby reduced from $N^2 = 81$ for direct evaluation to just 36.

Finally in this chapter, we consider the computation of the DFT of a real sequence. While the FFT algorithms previously developed were designed to accept a complex input sequence, and there are advantages to so doing, the use of such an algorithm with real data is inherently inefficient. One commonly-used possibility is to specialise the general N-point FFT computational structure to accept a length-N array of real values and to output a length-N array of $(N/2)$ complex values. This approach exploits the symmetry of the DFT spectrum of a real sequence about the folding frequency, f_N, which means that only those spectral values in the range $0 \leq f \leq f_N = f_s/2$ need be explicitly retained.

Given a general N-point algorithm capable of accepting complex input data, however, it can be used to transform efficiently a length-$2N$ real data sequence, $x(n)$, with only a small amount of additional computation. First, a 'complex' input sequence $g(n)$ is formed by making the 'real' sequence, $u(n)$, the even-indexed values of $x(n)$ and making the 'imaginary' sequence, $v(n)$, the odd-indexed values of $x(n)$. We then use the FFT to give $G(m) = \mathcal{D}[g(n)]$. From $G(m)$, the DFTs of $u(n)$ and $v(n)$ ($U(m)$ and $V(m)$ respectively) can be found very simply using the complex-conjugate (symmetry) property. This involves N complex multiplications. The desired transform $X(m)$ can be found in turn from $U(m)$ and $V(m)$, requiring another N complex multiplications, again exploiting the periodicity and symmetry properties. In total, an additional $2N$ complex multiplications are required, but for this modest cost we double the length of the input sequence which is transformed.

7.6 EXERCISES

7.1 Review the DFT properties studied in section 5.5 for the case of $x(n)$ real. How are these altered when $x(n)$ is complex?

7.2 In developing the decimation-in-time algorithm, the equality $W_N^2 = W_{N/2}$ was used (equation (7.3)). For the case of $N = 4$, plot the Nth roots of unity, W_N, on the complex plane and construct their squares, W_N^2. Confirm that these are equal to $W_{N/2}$.

Similarly, confirm the equality $W_N^{r+N/2} = -W_N^r$ (equation (7.7) which we used to reduce the number of complex multiplications for the FFT butterfly from two to one) for the case of $N = 4$ and for suitably-chosen values of r.

7.3 By amalgamating figures 7.1, 7.2 and 7.3, draw out the full signal flow graph for the 8-point, radix-2 decimation-in-time FFT and, hence, confirm that the number of complex multiplications is $N \log_2 N = 24$. Also draw out the graph for the 'butterfly computation' version in which the number of complex multiplications is reduced to $(N/2) \log_2 N = 12$.

7.4 The key to implementing the DFT efficiently is to exploit its periodicity and symmetry properties, which in turn derive from the periodicity and symmetry properties of the phase factors, W_N^{mn}. For the case of $N = 8$, evaluate and tabulate all $8 \times 8 = 64$ phase factors as m and n range from 0 to 8. How many distinct values are there? How many of these are real and how many are complex? How many pairs are related by sign reversal? How many pairs are complex conjugates? What are the implications for efficient implementations of the DFT?

7.5 Determine and tabulate the number of complex multiplications in (i) the radix-2 decimation-in-time FFT and (ii) the direct evaluation of the DFT, as the length of the transform, N, increases from 32 to 2048 in steps of integer powers of 2.

7.6 Draw the full signal flow graph for the $N = 8$ decimation-in-frequency algorithm and confirm that the number of complex multiplications required is $(N/2) \log_2 N = 12$.

7.7 Confirm the duality between decimation-in-time and decimation-in-frequency described on page 218 for a 4-point transform and its inverse.

7.8 From the signal flow graph for the 8-point, complex-input, decimation-in-time FFT obtained in exercise 7.3, deduce the simplifications which can be made when the input sequence is real. What is the effect on computational complexity in terms of the number of real multiplications?

8

Random signals

In the previous chapter, we introduced the idea of a deterministic signal, i.e. one which could be described by a mathematical formula which models the signal's generation process. It is almost exclusively this sort of signal that we have considered so far. Given previous values of a deterministic signal, it is possible correctly to predict (or determine) future values by fitting the observed values to the formula, hence the name. We have found very useful descriptions of deterministic DT signals to be the z-transform (chapter 3) and the discrete Fourier transform (chapter 5). It is a matter of experience, however, that not all signals are of this type. An example of a non-deterministic signal is the random noise component that was added to the sine wave of figure 2.1(b). It is the very essence of a random – or *stochastic* – signal that future values cannot be predicted by fitting observed, past values to a mathematical formula because the signal's generation process is too complex and/or poorly understood. Random signals occur commonly in science and engineering and are consequently of great importance. Examples include thermal noise generated in resistors in electronic circuits, speech sounds ('fricatives') generated by turbulent air flow in the vocal tract, and many forms of mechanical vibration in structures. A relevant question, therefore, is this: is it meaningful to describe such signals using the sort of techniques (e.g. the DFT spectrum, for instance) which we have found to be so useful in the case of deterministic signals? The purpose of our final chapter is to address this question: we will assume real signals throughout.

In order for the z-transform of a signal to exist, it must converge at all points of interest. Now, a (causal) limited-duration sequence $x(n)$ has a limited-duration ZT (of the form $\ldots + x(k-1)z^{-(k-1)} + x(k)z^{-k} + x(k+1)z^{-(k+1)} + \ldots$). Hence, such a sequence converges everywhere (except $z = 0$). For the DFT of a signal $x(n)$ to exist, it must be absolutely summable. While this is not the same as having finite energy, the latter (square summability) condition is actually weaker. In principle then, as we know from earlier chapters, there is no problem in using the z-transform or the DFT as a descriptor of a limited-duration, finite-energy signal. Nor is there any problem with a periodic, finite-power (infinite-energy) signal since its periodicity means that it is adequate to describe the signal over a single period (or an integer number of periods) – thereby treating it as effectively of limited duration. There is, however, a difficulty with the general class of random signals since these are both aperiodic and – because their length is indeterminate, so that they are in principle best considered to extend over all time – infinite-energy. Our purpose in this chapter is to study how this important class of signals can be included in the framework we have laid so far.

As we will see, the key is to describe random signals in terms of their statistical, average properties. These can never be a full description, in the way that the mathematical formula describing a deterministic signal is, but they can nonetheless be of value. In particular, it is sometimes possible to describe the random signal's statistical properties in a form which can be viewed as a finite-energy signal, so allowing our powerful Fourier techniques to be used. Unfortunately, however, exact determination of average properties is impractical, requiring an infinite observation interval. This reflects the statistical notion of probability as the limiting value of the ratio of actual, observed occurrences to possible occurrences, as the latter tends to infinity. Consequently, the values that we work with when dealing with random signals are always estimates, whose quality depends critically on statistical sampling effects, such as the available sequence length.

8.1 Moments of a random signal

Consider a random process, such as the generation of a thermal noise voltage by a resistor. Suppose we observe this voltage over some interval, and sample it to yield a DT signal. As this is one occasion where it is necessary to distinguish carefully between a sequence and a particular value of that sequence (see page 4 of chapter 1), we will denote this signal as $\{x(n)\}$. If we now repeat the observation, we will obtain a different $\{x(n)\}$. Each such $\{x(n)\}$ is then a single realisation, or *sample sequence*, of the random process. The entire set of such realisations is called the *ensemble* of the process. The reader is warned that some authors reserve the notation $\{x(n)\}$ for the ensemble rather than a single realisation. To avoid our notation becoming over-cumbersome, and in the belief that no confusion results, we will use $x(n)$ as a descriptor of the random process – as well as a particular value taken from a single realisation.

If we can describe the ensemble by some probability law then, although a particular

$\{x(n)\}$ will not be predictable from its past values, its average properties will be. In this way, we can partially yet usefully characterise the random signal.

A very obvious statistical, average descriptor of a signal is its mean value. The mean of the ensemble is defined as:

$$\mu = \lim_{N\to\infty} \frac{1}{N} \sum_{n=-N/2}^{N/2-1} x(n) \tag{8.1}$$

That is, it is the mean value of a single realisation as the length of that realisation tends to infinity and so becomes identical to the ensemble. This is sometimes called the *expected value* or *expectation*, $E()$, of $x(n)$ so that:

$$E(x_n) = \mu \tag{8.2}$$

It is sometimes helpful when we are dealing with multiple processes to distinguish the particular random signal described by μ. In such a case, the above value is denoted μ_x.

The mean of the ensemble can be estimated from a sample sequence of length N as the *sample mean*:

$$\hat{\mu} = \frac{1}{N} \sum_{n=0}^{N-1} x(n) \approx \mu \tag{8.3}$$

where the 'hat', $\hat{}$, is used here and subsequently to indicate a statistical estimation. For consistency with the usual notation, note that we have let the index n run from 0 to $(N-1)$.

Suppose $x(n)$ adopts one of k discrete values $y_1, y_2, \ldots, y_k$, and does so with probabilities $p_1, p_2, \ldots, p_k$ respectively. Then in a sequence of length N tending to infinity, we will observe y_1 Np_1 times, y_2 Np_2 times etc. Hence, equation (8.1) becomes:

$$\begin{aligned} \mu &= \lim_{N\to\infty} \frac{1}{N}(Np_1y_1 + Np_2y_2 + \cdots + Np_ky_k) \\ &= \sum_{m=1}^{k} y_k p_k \end{aligned}$$

In the general case that the values of $x(n)$ are continuous, this becomes:

$$\mu = \int_{-\infty}^{\infty} yp(x_n, y)dy \tag{8.4}$$

Here, $p(x_n, y)$ is the *probability density function* which we will study in some detail below, where we will give a careful interpretation to the dummy variable y.

The ensemble mean describes the random signal's so-called central tendency. Another useful signal descriptor is its average fluctuation. Since this can be negative as well as positive, we take the expected value of $x^2(n)$:

$$E(x_n^2) = \lim_{N\to\infty} \frac{1}{N} \sum_{n=-N/2}^{N/2-1} x^2(n) \tag{8.5}$$

Comparing with equation (1.13), we see that this is identically equal to the average power of $\{x(n)\}$.

It is worth noting at this stage that the expectation of a product is not generally equal to the product of the individual expected values:

$$E(x_n y_n) \neq E(x_n)E(y_n) = \mu_x \mu_y$$

If this were so, then $E(x_n^2)$ would be equal to μ^2 rather than to the average power. The equality does, however, hold when the random processes $\{x(n)\}$ and $\{y(n)\}$ are *statistically independent*. Clearly, $\{x(n)\}$ cannot be independent of itself!

Generalising equations (8.1), (8.2) and (8.5) to the case of any power l:

$$E(x_n^l) = \lim_{N\to\infty} \frac{1}{N} \sum_{n=-N/2}^{N/2-1} x^l(n) \tag{8.6}$$

$E(x_n^l)$ is called the lth *moment* of the ensemble. Most often, only the first (mean, $l = 1$) and second (average fluctuation, $l = 2$) moments are employed as signal descriptors, but the higher moments are occasionally useful.

Proceeding as before in going from equation (8.1) to equation (8.4), when the possible values adopted by $x(n)$ are continuous, the lth moment can be found in terms of the probability density function as:

$$E(x_n^l) = \int_{-\infty}^{\infty} y^l p(x_n, y)dy \tag{8.7}$$

It is also possible to define *central* moments of the ensemble. These moments are taken around the mean, i.e. they are the expected values of $(x_n - \mu)^l$. Obviously, the first central moment is uninteresting, being trivially equal to zero. By contrast, following equation (8.5), the second central moment:

$$\begin{aligned} E\left[(x_n - \mu)^2\right] &= \overline{(x_n - \mu)^2} \\ &= \lim_{N\to\infty} \frac{1}{N} \sum_{n=-N/2}^{N/2-1} (x_n - \mu)^2 \end{aligned}$$

has an interpretation as the average power about the mean, or the 'ac' power. The ensemble value of the second central moment is also known as the *variance*, σ^2. Again, we use subscripting, σ_x^2, if it is necessary to specify which of a number of random process is being considered.

From a particular realisation, $\{x(n)\}$, σ can be estimated as the standard deviation of the sample:

$$\hat{\sigma} = \frac{1}{N-1}\sqrt{\sum_{n=0}^{N-1}(x_n - \hat{\mu})^2} \approx \sigma \tag{8.8}$$

The normalisation here by $(N-1)$, rather than by N, renders $\hat{\sigma}$ an *unbiased estimator* of σ (see below and section 8.8). The distinction is only of practical importance for small N.

8.2 STATIONARITY AND ERGODICITY

Equations (8.3) and (8.8) specify how the mean and variance of a random process can be estimated from a single sample sequence $\{x(n)\}$ of the process by averaging over time, according to the index n. Implicit in our development in the previous section was the assumption that the moments of the random process are not functions of time: rather, they take constant values independent of time. Hence, averaging over time using a sample sequence will give estimates of these constant values. The ensemble values themselves were defined as averages over *all* time.

There is another way to obtain average statistical values which uses multiple realisations as depicted in figure 8.1. Here, K realisations have been aligned and averages taken at a particular time index n_0 across all sequences; a process called *ensemble averaging*. Clearly, as $K \to \infty$, the obtained averages will tend to the ensemble values. A random process for which the ensemble average values are independent of the time index n_0 at which they were computed is called *stationary*. For such a process, its average properties computed for a finite number of realisations, K, do not change with time index n_0 other than to exhibit random variation due to statistical sampling. In fact, a sequence of estimates of the ensemble moments produced by some estimator as n_0 varies will itself be a random process. If the expected (mean) value is equal to the moment, or ensemble value, itself then we say that the estimator is *unbiased*. To illustrate this, suppose we have a number of estimates, $\hat{\mu}$, of the ensemble mean, μ obtained using equation (8.3). Then:

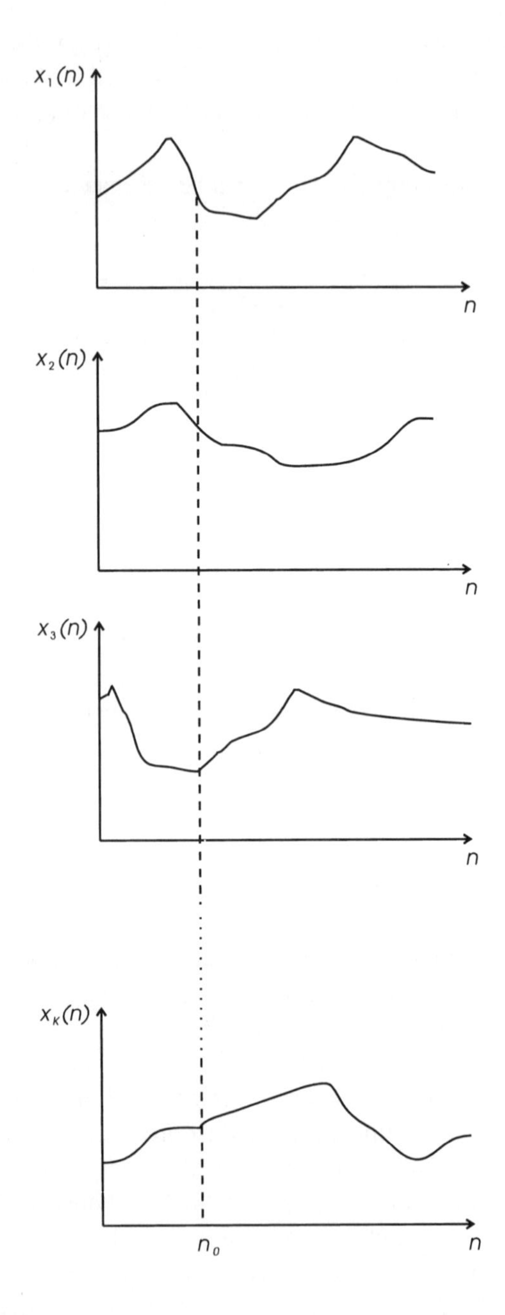

Fig. 8.1 As an alternative to averaging over time, statistical moments of a random signal can be obtained by ensemble averaging.

$$
\begin{aligned}
E[\hat{\mu}] &= E\left[\frac{1}{N}\sum_{n=0}^{N-1} x(n)\right] \\
&= \frac{1}{N}\sum_{n=0}^{N-1} E[x_n] \\
&= \frac{1}{N} N\mu \\
&= \mu
\end{aligned}
$$

and the estimator is unbiased.

A stationary random process for which the moments obtained by ensemble averaging equal those obtained by time averaging is called *ergodic*. Clearly, ergodicity is a stronger condition than stationarity. Ergodic processes are of considerable practical importance, since they can be characterised approximately by averages taken over a single realisation: the longer the sample sequence, the better is the approximation. Throughout this chapter, we will assume the property of ergodicity holds.

8.3 THE PROBABILITY DENSITY FUNCTION

The general nature of equation (8.7) suggests that the probability density function itself could be a very useful signal descriptor. Can we characterise the ensemble by the probability distribution of the $x(n)$ values? If (as we have assumed throughout) these are continuous, it makes no sense to consider the probability of $x(n)$ being precisely equal to some exact value, y, since this is necessarily zero. Rather, we consider the probability that $x(n)$ lies in some narrow range of values of width δy about y. Treating probability as the limiting value of relative frequency of occurrence:

$$
\text{prob}\left[\left(y - \frac{\delta y}{2}\right) < x(n) < \left(y + \frac{\delta y}{2}\right)\right] = \lim_{N\to\infty}\left(\frac{N(x_n, y, \delta y)}{N}\right) \tag{8.9}
$$

where $N(x_n, y, \delta y)$ is the number of $x(n)$ values observed in the range $y \pm (\delta y/2)$ and N is the total number of observations, i.e. the length of the sample sequence.

The basis of our difficulty is that, as δy tends to zero, so too does this probability. However, the ratio of this probability to δy tends to a real, finite value which is not identically zero. This ratio defines the probability density function (PDF):

$$
p(x_n, y) = \lim_{\delta y\to 0}\lim_{N\to\infty}\left(\frac{N(x_n, y, \delta y)}{N\delta y}\right) \tag{8.10}
$$

which we have already come across above. The concept of *density* here is very similar to that introduced in section 1.9.2 for the energy density spectrum of a (continuous-time) aperiodic signal.

It follows from this development, whereby $p(x_n, y)$ is defined in terms of the limiting value of the probability of $x(n)$ lying in a range of values of width δy, that the probability of $x(n)$ lying between x_1 and x_2 is:

$$\text{prob}[x_1 < x(n) < x_2] = \int_{x_1}^{x_2} p(x_n, y)dy \tag{8.11}$$

and, hence, the integral of the PDF is 1 over all possible values of $x(n)$:

$$\int_{-\infty}^{\infty} p(x_n, y)dy = 1 \tag{8.12}$$

Often, it is not necessary to refer to the dummy variable y which can consequently be omitted, and the PDF written more simply as $p(x_n)$. The reader is warned not to interpret this notation as indicating that $p(x_n)$ is a probability: as we have been at pains to point out, it is a probability density. Keeping this distinction clear is the reason we have used prob( ) to denote a probability, rather than $p()$.

Given that y is a dummy variable, it follows from the general equation (8.7) that the PDF embodies all the information about the signal contained in its l moments. Note that, unlike a probability, the PDF is not upper bounded by +1. If we consider, for instance, a 'random' sequence which is identically zero, then its PDF will be a Dirac delta function, $\delta(x_n)$, which is unbounded at $x(n) = 0$. On the other hand, the PDF is lower bounded by 0, i.e. it is a non-negative function. This follows from the fact that the probability in equation (8.9) is bounded by 0 and 1, while δy in equation (8.10) is non-negative.

8.3.1 The rectangular PDF

A PDF commonly encountered is the rectangular distribution:

$$p(x_n) = \begin{cases} P & L \leq x(n) \leq U \\ 0 & \text{otherwise} \end{cases} \tag{8.13}$$

where P is a constant, and L and U are the lower and upper bounds respectively on the amplitude of $x(n)$. As the concept of density emphasises, this PDF must not be interpreted as indicating that the probability of $x(n)$ taking any value between L and U is P. Rather, the probability of $x(n)$ being within any particular range $y \pm (\delta y/2)$ is the

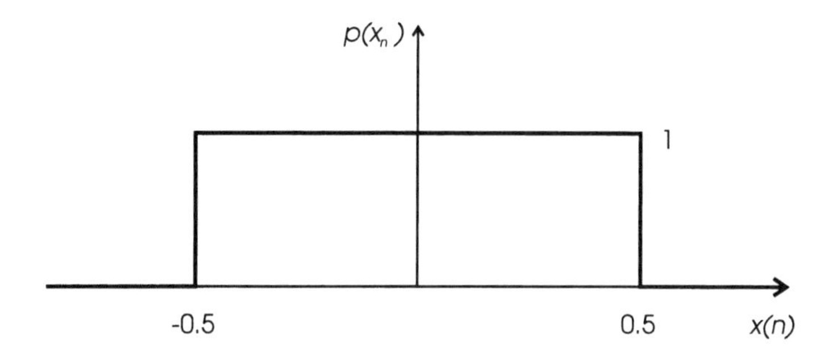

Fig. 8.2 Rectangular probability distribution function for a random process with $L = -0.5$ and $U = 0.5$.

same as it being within any other range of width δy bounded by L and U, independent of the value of y.

Since the total area under the PDF must be 1 (equation (8.12)), it is clear that:

$$P = \frac{1}{U - L} \tag{8.14}$$

Figure 8.2 depicts the rectangular PDF with $L = -0.5$ and $U = 0.5$. Hence, $P = 1$ in this case.

In the practical case of non-zero δy and finite N, the PDF can be estimated from a sample sequence by simplifying equation (8.10) as:

$$\hat{p}(x_n) \approx \frac{N(x_n, y, \delta y)}{N \delta y} \tag{8.15}$$

As an illustration, figure 8.3(a) shows a 200-point sample sequence generated from the random process with PDF depicted in figure 8.2. From equation (8.7) with $l = 1$, the ensemble mean is:

$$\begin{aligned} \mu &= \int_{-\infty}^{\infty} y p(x_n, y) dy \\ &= \int_{-0.5}^{0.5} y dy \\ &= \left[\frac{y^2}{2} \right]_{-0.5}^{0.5} \end{aligned}$$

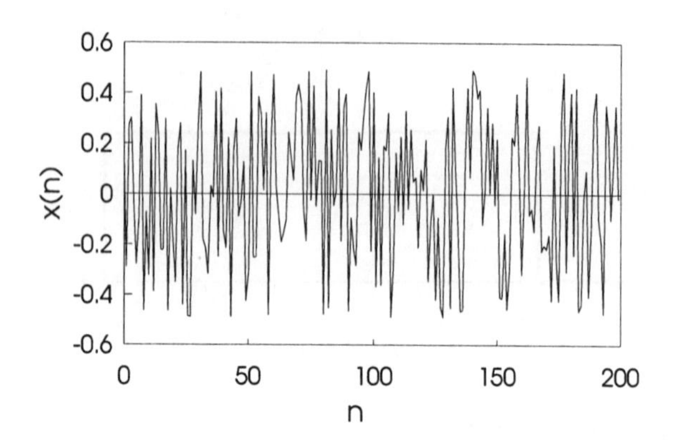

(a)

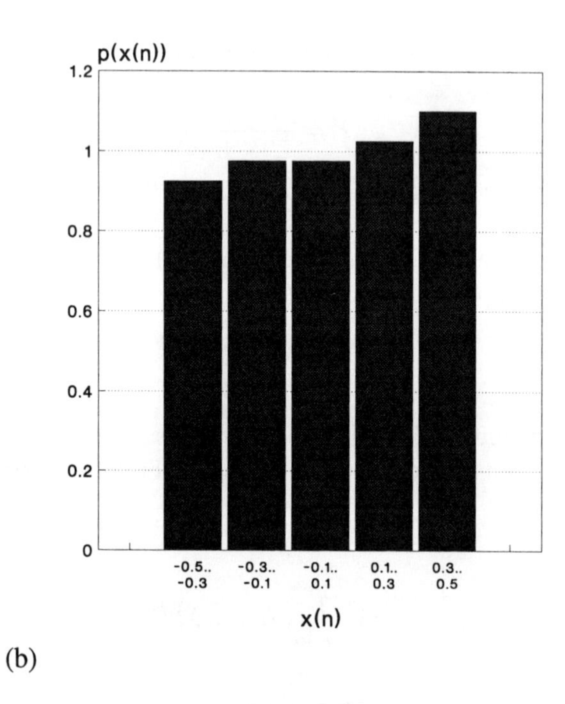

(b)

Fig. 8.3 (a) 200-point sample sequence generated from a random process with rectangular probability distribution function between ± 0.5 and (b) PDF estimated from this 200-point sequence with $\delta y = 0.2$.

$$= 0$$

Since the mean is zero, the second moment and the second central moment (variance) are identical. Hence, from equation (8.7) with $l = 2$:

$$\begin{aligned}
\sigma^2 &= \int_{-\infty}^{\infty} y^2 p(x_n, y) dy \\
&= \int_{-0.5}^{0.5} y^2 dy \\
&= \left[\frac{y^3}{3} \right]_{-0.5}^{0.5} \\
&= 0.0833
\end{aligned}$$

Figure 8.3(b) shows the estimated PDF for this particular realisation obtained from the sample sequence using equation (8.15) with $\delta y = 0.2$ and $N = 200$. That is, the number of $x(n)$ values within each range of width 0.2 was counted and then divided by $N\delta y = 40$. The estimated function is approximately flat (rectangular) and the area under it is 1. If it accurately reflected the PDF of the generating process, however, there would be exactly $200/5 = 40$ $x(n)$ values within each bin of width $\delta y = 0.2$ and a $\hat{p}(x_n)$ value of 40/40 = 1.0 in all cases. However, the obtained values actually range from 0.925 to 1.1. Also, the sample mean is calculated as $\hat{\mu} = 0.0101$ (rather than zero) while the sample variance is $\hat{\sigma}^2 = 0.0876$ (rather than 0.0833). These discrepancies are, of course, a consequence of the random (stochastic) nature of the generation process which means that (for an ergodic process at least) it is necessary to take the limiting case of an infinitely long observation interval. Also, according to equation (8.10), δy should in principle be infinitesimally small (although this is less important in the case of a rectangular distribution.)

Accordingly, we can improve our estimates of the ensemble properties of an ergodic random process in a statistical sense by increasing the observed length, N, of the sample sequence. Figure 8.4 shows the estimated PDF when the length of the sample sequence is increased from 200 points to 2000. While the function is only very slightly flatter, with the observed values ranging from 0.93 to 1.09 (c.f. values of 0.925 and 1.1 for the 200-point sequence), the sample mean is now $\hat{\mu} = 0.0045$ and the sample variance is $\hat{\sigma}^2 = 0.0837$ – significantly closer to the ensemble values $\mu = 0$ and $\sigma^2 = 0.0833$ respectively.

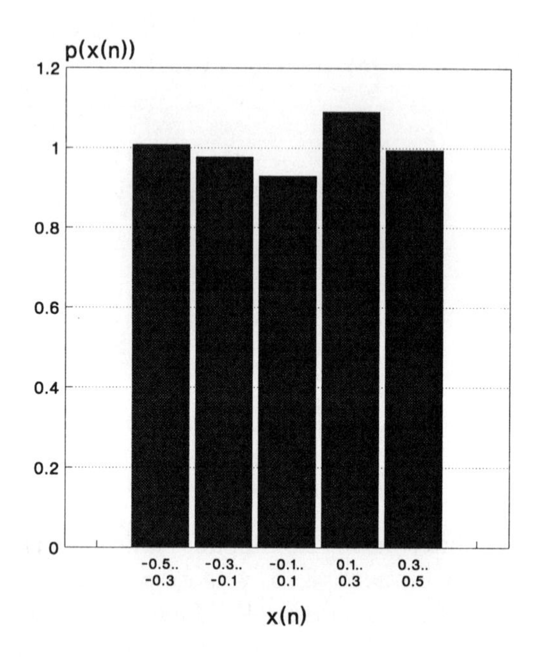

Fig. 8.4 Estimated PDF when the sample sequence length is increased from 200 to 2000 points.

8.3.2 The PDF of a sine wave

As an example illustrating the use of the rectangular distribution, let us find the PDF of the (zero-mean) DT sine wave with M samples per period:

$$x(n) = A\sin\left(\frac{2\pi n}{M} + \theta\right) \tag{8.16}$$

This, however, is a deterministic signal: yet there seems no reason why we should not describe it by statistical moments such as its mean and variance. How then can it be cast as a random signal?

Following our discussion of ensemble averaging, we consider K different realisations of $x(n)$ randomly sampled over time, and fix the time index at $n = n_0$. The effect is to make the phase angle $\theta(k)$ for the kth realisation at time index n_0 a random variable. We can then express the sine wave as a function of this random variable thus:

$$x(\theta_k) = A \sin\left(\frac{2\pi n_0}{M} + \theta(k)\right)$$

Now, it is reasonable to suppose that the random variable $\theta(k)$ is rectangularly distributed between 0 and 2π radians. Hence, using equations (8.13) and (8.14):

$$p(\theta_k) = \begin{cases} 1/2\pi & 0 \le \theta(k) \le 2\pi \\ 0 & \text{otherwise} \end{cases} \tag{8.17}$$

So we know $p(\theta_k)$, but we want $p(x_n)$. From equations (8.9) and (8.10), replacing the dummy value y by x, $p(x_n)$ is defined as the limiting value:

$$p(x_n) = \lim_{\delta x \to 0} \left(\frac{\text{prob}\left[\left(x - \frac{\delta x}{2}\right) < x(n) < \left(x + \frac{\delta x}{2}\right)\right]}{\delta x} \right)$$

and, similarly:

$$p(\theta_k) = \lim_{\delta \theta \to 0} \left(\frac{\text{prob}\left[\left(\theta - \frac{\delta \theta}{2}\right) < \theta(k) < \left(\theta + \frac{\delta \theta}{2}\right)\right]}{\delta \theta} \right)$$

Because $x(n)$ has a sinusoidal dependence on $\theta(k)$, $x(n)$ can take one of two values for a given $\theta(k)$ in the range $(0,2\pi)$ while $\theta(k)$ itself takes only 1. Therefore, the probability of $x(n)$ lying in a range $x \pm (\delta x/2)$ is – provided we normalise by δx which is small – twice the probability of $\theta(k)$ lying in a range $\theta \pm (\delta\theta/2)$:

$$\frac{\text{prob}\left[\left(x - \frac{\delta x}{2}\right) < x(n) < \left(x + \frac{\delta x}{2}\right)\right]}{\delta x} = 2 \frac{\text{prob}\left[\left(\theta - \frac{\delta \theta}{2}\right) < \theta(k) < \left(\theta + \frac{\delta \theta}{2}\right)\right]}{\delta x}$$

Hence:

$$\begin{aligned} p(x_n) &= 2 \lim_{\delta x \to 0} \left(\frac{\text{prob}\left[\left(\theta - \frac{\delta \theta}{2}\right) < \theta(k) < \left(\theta + \frac{\delta \theta}{2}\right)\right]}{\delta x} \right) \\ &= 2 \lim_{\delta x \to 0} \lim_{\delta \theta \to 0} \frac{p(\theta_k)\delta\theta}{\delta x} \\ &= 2 p(\theta_k) \frac{d\theta}{dx} \end{aligned}$$

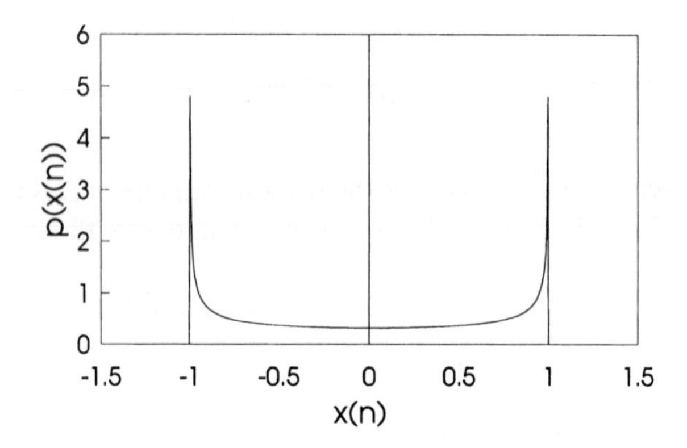

Fig. 8.5 Probability density function of sine wave of amplitude $A = 1$.

But:

$$\begin{aligned}
\frac{dx}{d\theta} &= A\cos\left(\frac{2\pi n_0}{M} + \theta\right) \\
&= A\sqrt{1 - \sin^2\left(\frac{2\pi n_0}{M} + \theta\right)} \\
&= A\sqrt{1 - x^2(n)} \\
&= \sqrt{A^2 - x^2(n)} \\
\text{and} \quad p(x_n) &= \frac{2p(\theta_k)}{\sqrt{A^2 - x^2(n)}}
\end{aligned}$$

Substituting into equation (8.17):

$$p(x_n) = \begin{cases} 1/\left(\pi\sqrt{A^2 - x^2(n)}\right) & -A \leq x(n) \leq A \\ 0 & \text{otherwise} \end{cases} \tag{8.18}$$

Figure 8.5 shows this PDF for a sine wave with $A = 1$. Note that this is an instance where $p(x_n)$ is unbounded, specifically at $x(n) = \pm A$ (see exercise 8.3(b)). The example illustrates how the PDF can only partially characterise a random process. The

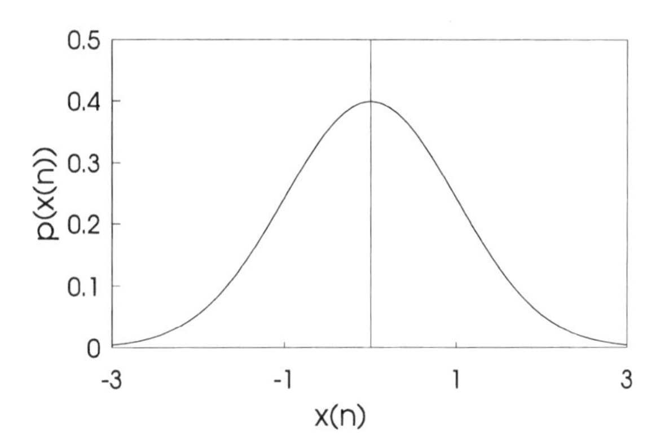

Fig. 8.6 Probability density function of Gaussian random process $N(0, 1)$ having zero mean and unit variance.

figure contains no information about the phase of the sine wave, considered as a 'random' signal, nor even its frequency, only about its average properties. Hence, an infinite number of different signals can have the same PDF which, therefore, defines a class of random signals. By contrast, given sufficient previous values (just 3, as only the frequency, amplitude and phase need be determined) of equation (8.16), we can predict any future value. Of course, this is not an option for a true random signal where no such deterministic equation exists.

8.3.3 The Gaussian PDF

Another very important class of random process is that having a Gaussian, or *normal* PDF:

$$p(x_n) = \frac{1}{\sigma\sqrt{2\pi}} \exp\left(\frac{-(x_n - \mu)^2}{2\sigma^2}\right) \tag{8.19}$$

This is illustrated in figure 8.6 for the particular case of zero mean and unit variance. For such a process, $x(n)$ is continuous and unbounded: it can, in principle, take any value between $\pm\infty$. The PDF is symmetrical and bell-shaped, peaking at the mean value μ. The width of the peak reduces as σ decreases, while the peak value increases to keep the area under the PDF constant at 1. A normal random process is often characterised as $N(\mu, \sigma)$: the PDF in figure 8.6 describes $N(0, 1)$.

Figure 8.7(a) shows a 200-point sample sequence generated from a normal random process $N(0, 1)$. The sample mean and variance are calculated as -0.0055 and 0.9850

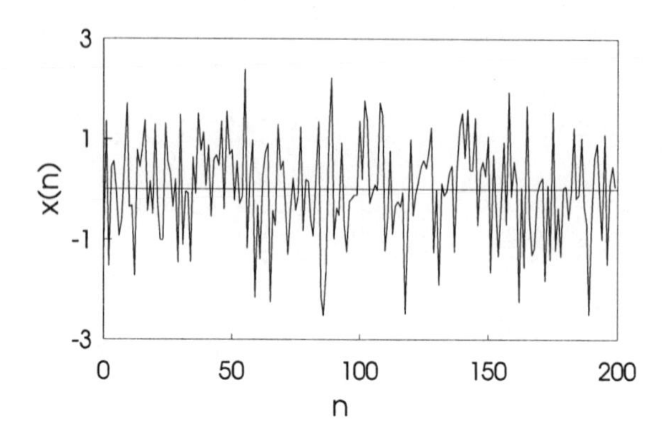

(a)

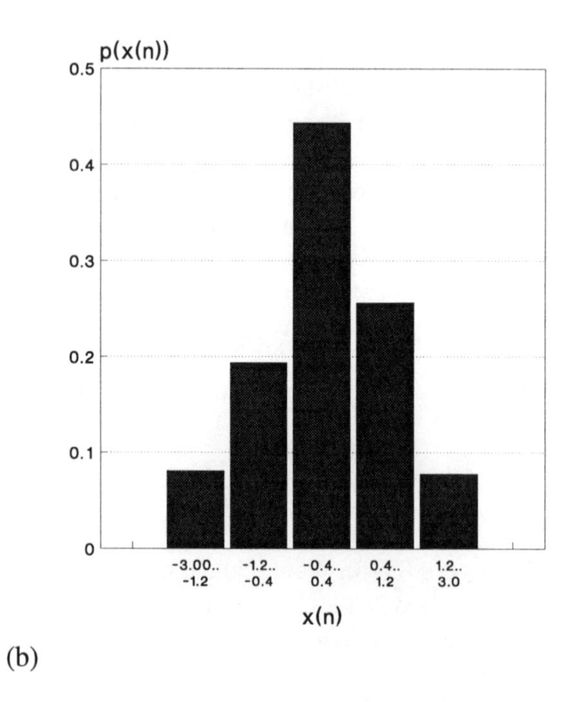

(b)

Fig. 8.7 (a) 200-point sample sequence generated from a Gaussian random process $N(0, 1)$ with zero mean and unit variance and (b) PDF estimated from this 200-point sequence.

respectively as opposed to 0 and 1. Figure 8.7(b) depicts the PDF estimated from the sample sequence using equation (8.15) with $\delta y = 0.8$. Because $x(n)$ is unbounded, however, the bins at the ends of the distribution have been widened to $\delta y = 1.8$ to encompass the extreme values of the sequence.

Given that N is finite and δy is not infinitesimally small, how does this estimated PDF compare with the actual PDF of the process? From equations (8.11) and (8.19), the probability that a particular $x(n)$ lies in the range 0 to X can be obtained as:

$$\text{prob}(0 \le x_n \le X) = \frac{1}{\sigma\sqrt{2\pi}} \int_0^X \exp\left(\frac{-(y-\mu)^2}{2\sigma^2}\right) dy$$

with values for $N(0, 1)$ tabulated against X in many texts on elementary statistics. From such a tabulation, denoting $p(0 \le x_n \le X)$ as $P(X)$ and exploiting the symmetry of the Gaussian PDF, we find:

$$\begin{aligned}
\text{prob}(-3 < x_n \le -1.2) &= P(3) - P(1.2) &&= 0.1138 \\
\text{prob}(-1.2 < x_n \le -0.4) &= P(1.2) - P(0.4) &&= 0.2294 \\
\text{prob}(-0.4 < x_n \le 0.4) &= P(0.4) + P(0.4) &&= 0.3110 \\
\text{prob}(0.4 < x_n \le 1.2) &= P(1.2) - P(0.4) &&= 0.2294 \\
\text{prob}(1.2 < x_n \le 3) &= P(3) - P(1.2) &&= 0.1138
\end{aligned}$$

Note that these do not quite sum to 1 because of the finite but small probability that $x(n)$ lies outside the range $(-3, +3)$. With $N = 200$, we expect on average (over repeated trials) to count 22.76, 45.88, 62.2, 45.88 and 22.76 $x(n)$ values within these 5 bins respectively. In fact, the actual values obtained are 29, 31, 70, 42 and 28.

8.4 THE AUTOCORRELATION FUNCTION

The autocorrelation function (ACF) of a random signal $\{x(n)\}$ describes the way that values at one time relate to values at some other time. It is a real function defined as the expectation:

$$E[x(n)x(n+\tau)] = R_{xx}(\tau) = \lim_{N\to\infty} \frac{1}{N} \sum_{n=-N/2}^{N/2-1} x(n)x(n+\tau) \tag{8.20}$$

where τ represents a time shift by an integer number of sample periods. By defining the ACF as a function of the shift parameter τ, thereby embodying time information in the statistical average, we hope to portray more information about the process than do any of the individual moments or, indeed, the PDF.

As with the other expected values of an ergodic random process, an estimate of this function can be obtained by time averaging over a realisation of finite length N. We will deal with this issue in section 8.8 below. In the meantime, we assume that infinite-duration sequences are available.

Taking $x(n)$ to be a continuous variable, we can (as we did in going from equation (8.1) to equation (8.4) earlier) define the ACF in integral form with dummy variable y as:

$$R_{xx}(\tau) = \int_{-\infty}^{\infty} y(n)y(n+\tau)p(x_n, y)dy \tag{8.21}$$

By substituting $k = (n+\tau)$ for n in equation (8.20), we can express the ACF in the form of a convolution-sum:

$$R_{xx}(\tau) = \lim_{N\to\infty} \sum_{k=-N/2+\tau}^{N/2-1+\tau} x(k-\tau)x(k)$$

By ignoring τ in relation to $N/2$ as $N \to \infty$, and using the commutativity property of the convolution-sum, we obtain:

$$\begin{aligned} R_{xx}(\tau) &= \lim_{N\to\infty} \sum_{k=-N/2}^{N/2-1} x(k)x(k-\tau) \\ &= R_{xx}(-\tau) \end{aligned}$$

so that $R_{xx}(\tau)$ is an even function.

Setting τ equal to 0 in equation (8.20):

$$R_{xx}(0) = E[x^2(n)]$$

which, according to equation (8.5), is the average power of the sequence. This is the maximum value that $R_{xx}(\tau)$ can reach, since the maximum obtains only when $x(n)$ and $x(n+\tau)$ are identical. In fact, the value is an extremum since:

$$|R_{xx}(\tau)| \leq R_{xx}(0)$$

Consider a zero-mean random sequence for which adjacent values are statistically independent. In this case, $x(n)$ and $x(n+\tau)$ will likewise be independent and, as outlined on page 230, the expected value of their product will be equal to the product of the two expected values:

$$
\begin{aligned}
R_{xx}(\tau) &= E[x(n)x(n+\tau)] \\
&= E[x(n)]E[x(n+\tau)] \\
&= \begin{cases} E[x^2(n)] & \tau = 0 \\ \mu^2 = 0 & \tau \neq 0 \end{cases}
\end{aligned}
\tag{8.22}
$$

so that, unlike any of the lth moments, the ACF embodies information about both the mean and average power (in this case) which is separately recoverable.

Equation (8.22) reveals that the ACF of random noise, for which adjacent values are independent and with zero mean, is a (scaled) Kronecker delta function of amplitude equal to the average noise power at $\tau = 0$. This is a result of some importance: recall that our purpose in this chapter is to characterise infinite-duration, infinite-energy random signals. Here we have characterised such a signal by a very compact (just one value!) sequence indeed.

If the sequence has adjacent values statistically independent, but non-zero mean then, from equation (8.22):

$$\mu = \sqrt{R_{xx}(\tau)} \qquad \tau \neq 0 \tag{8.23}$$

A signal measure related to the ACF is the *autocovariance function*, defined (for a stationary random process) as:

$$
\begin{aligned}
\gamma_{xx}(\tau) &= E[(x_n - \mu)(x_{n+\tau} - \mu)] \\
&= E[(x_n)(x_{n+\tau}) - (x_n)\mu - (x_{n+\tau})\mu + \mu^2] \\
&= E[(x_n)(x_{n+\tau})] - E(x_n)\mu - E(x_{n+\tau})\mu + \mu^2 \\
&= R_{xx}(\tau) - \mu^2 - \mu^2 + \mu^2 \\
&= R_{xx}(\tau) - \mu^2
\end{aligned}
$$

It is easily shown (as a simple consequence of exercise 8.1) that $\gamma_{xx}(0) = \sigma^2$. Like the ACF, the autocovariance function is even and the value attained at $\tau = 0$ (equal to the variance) is an extremum:

$$|\gamma_{xx}(\tau)| \leq \gamma_{xx}(0)$$

In the case that adjacent values of $x(n)$ are independent then, from equation (8.23):

$$\gamma_{xx}(\tau) = 0 \qquad \tau \neq 0$$

If the autocovariance function is identically zero for all τ (except $\tau = 0$), we refer to adjacent $x(n)$ values as *uncorrelated.* This is a related – but not strictly identical – concept to statistical independence. For our purposes, however, we can treat them as synonyms.

If adjacent values of a random sequence are *partially correlated*, then the ACF will fall away (symmetrically) from its value of $E[x^2(n)]$ at $\tau = 0$ to a constant value of μ^2 as $|\tau|$ increases:

$$\mu = \sqrt{R_{xx}(\infty)}$$

and the rate of fall-off is an (inverse) indication of the degree of correlation. The faster the fall-off, the less is the degree of correlation.

This equality is satisfied generally for a random signal, since if a signal is fully correlated, then it is not random but deterministic. Let us illustrate this by finding the ACF of a sine wave and showing that its amplitude does not fall off at all.

$$x(n) = A \sin\left(\frac{2\pi n}{M} + \theta\right)$$

$$\text{and} \quad x(n+\tau) = A \sin\left(\frac{2\pi(n+\tau)}{M} + \theta\right)$$

Following section 8.3.2, we will consider both of these to be functions of random, rectangularly-distributed phase:

$$x(n) = x_1(\theta)$$

$$\text{and} \quad x(n+\tau) = x_2(\theta)$$

Now, the phase is rectangularly and continuously distributed between 0 and 2π according to equation (8.17). Because θ is continuous, we use the integral form of the ACF (equation (8.21)) with dummy variable θ:

$$R_{xx}(\tau) = \int_{-\infty}^{\infty} x_1(\theta) x_2(\theta) p(\theta) d\theta$$

$$= \frac{1}{2\pi}\int_0^{2\pi} x_1(\theta)x_2(\theta)d\theta$$

$$= \frac{A^2}{2\pi}\int_0^{2\pi} \sin\left(\frac{2\pi n}{M}+\theta\right)\sin\left(\frac{2\pi(n+\tau)}{M}+\theta\right)d\theta$$

Using the identity:

$$\sin a \sin b = \frac{1}{2}[\cos(a-b) - \cos(a+b)]$$

gives:

$$R_{xx}(\tau) = \frac{A^2}{2\pi}\left[\frac{1}{2}\int_0^{2\pi} \cos\left(\frac{2\pi\tau}{M}\right)d\theta - \frac{1}{2}\int_0^{2\pi} \cos\left(\frac{4\pi n}{M}+\frac{2\pi\tau}{M}+2\theta\right)d\theta\right]$$

Since the second integral is over an integer number (two) of cycles, it disappears, and:

$$\begin{aligned} R_{xx}(\tau) &= \frac{A^2}{4\pi}\cos\left(\frac{2\pi\tau}{M}\right)\int_0^{2\pi} d\theta \\ &= \frac{A^2}{4\pi}\cos\left(\frac{2\pi\tau}{M}\right)\Big[\theta\Big]_0^{2\pi} \\ &= \frac{A^2}{2}\cos\left(\frac{2\pi n\tau}{M}\right) \end{aligned} \qquad (8.24)$$

Thus, $R_{xx}(0) = A^2/2$ which is (as expected) equal to the average power of the sinusoidal signal. When $\tau \neq 0$, the $\cos(2\pi n\tau/M)$ term acts as a power factor. Also, the ACF (unlike the lth moments) retains important time-domain information about the signal. Clearly, the frequency of the sine wave can be recovered from $R_{xx}(\tau)$, although the phase cannot.

Having studied the autocorrelation function, we can now refine our definition of stationarity somewhat. If all probability distributions of a random process are independent of time, then the process is *strongly* or *strictly* stationary. If, however, the distributions are not generally time-independent, but the mean and ACF are, then the process is *weakly* stationary or stationary *in the wide sense.*

8.5 THE CROSS-CORRELATION FUNCTION

Thus far, we have considered a single random process, $x(n)$, whose PDF, $p(x_n)$, is defined as the limiting value:

$$p(x_n) = \lim_{\delta x \to 0} \left(\frac{\text{prob}\left[\left(x - \frac{\delta x}{2}\right) < x(n) < \left(x + \frac{\delta x}{2}\right)\right]}{\delta x} \right)$$

Often, we wish to consider the relation between two random processes, $x(n)$ and $y(n)$ say. In this case, we can characterise the relation by the joint statistics of the two processes. We extend the above definition to give the joint PDF as:

$$p(x_n, y_n) = \lim_{\delta x \to 0} \lim_{\delta y \to 0} \left(\frac{P}{\delta x \delta y} \right)$$

where:

$$P = \text{prob}\left[\left(x - \frac{\delta x}{2}\right) < x(n) < \left(x + \frac{\delta x}{2}\right) \text{ and } \left(y - \frac{\delta y}{2}\right) < y(n) < \left(y + \frac{\delta y}{2}\right)\right]$$

Thus, the joint PDF of $x(n)$ and $y(n)$ is a two-dimensional function. The notation $p(x_n, y_n)$ should not be confused with that for the one-dimensional PDF, $p(x_n, y)$, used earlier – where y was a dummy variable.

The expected value of the product $x(n)y(n + \tau)$ is of some interest. It is called the *cross-correlation function* (CCF), and is given as:

$$E[x(n)y(n + \tau)] = R_{xy}(\tau) = \lim_{N \to \infty} \frac{1}{N} \sum_{n=-N/2}^{N/2-1} x(n)y(n + \tau)$$

Following equation (8.21), when the $x(n)$ and $y(n)$ values are continuous, the CCF can be defined in terms of the joint PDF as:

$$R_{xy}(\tau) = \int_{-\infty}^{\infty} \int_{-\infty}^{\infty} x(n)y(n + \tau)p(x_n, y_n)dxdy$$

Clearly, this becomes identical to the ACF, $R_{xx}(\tau)$, when $y(n) = x(n)$.

The cross-correlation function yields important information about how $x(n)$ and $y(n)$ co-vary. Suppose, for example, that $\{x(n)\}$ and $\{y(n)\}$ are actually identical, zero-mean realisations apart from some time shift, Δ. This may be very difficult, if not impossible, to spot from the time waveforms. However, the cross-correlation will display a sharp peak, equal to the average power of $x(n)$ ($= y(n)$), at $\tau = \Delta$. If adjacent values of $x(n)$ (and, therefore, of $y(n)$) are statistically independent, the CCF will actually be a (scaled) Kronecker delta function, zero at all points except $\tau = \Delta$. It follows that, unlike the ACF, the CCF is not an even function. In fact (exercise 8.5):

$$R_{xy}(\tau) = R_{yx}(-\tau)$$

Next suppose that $x(n)$ and $y(n)$ are statistically independent. Then:

$$\begin{aligned} R_{xy}(\tau) &= E[x(n)y(n+\tau)] \\ &= E[x(n)]E[y(n+\tau)] \\ &= \mu_x \mu_y \end{aligned}$$

and the CCF will be constant for all τ.

As we did in the previous section, we can define a related function to the CCF called the *cross-covariance function*:

$$\begin{aligned} \gamma_{xy}(\tau) &= E[(x_n - \mu_x)(y_{n+\tau} - \mu_y)] \\ &= E[(x_n)(y_{n+\tau})] - E(x_n)\mu_y - E(y_{n+\tau})\mu_x + \mu_x\mu_y \\ &= \gamma_{xy}(\tau) - \mu_x\mu_y - \mu_y\mu_x + \mu_x\mu_y \\ &= \gamma_{xy}(\tau) - \mu_x\mu_y \end{aligned}$$

The cross-covariance function will then be a (scaled) Kronecker delta function at $\tau = \Delta$ when $x(n)$ and $y(n)$ are identical but shifted by Δ samples. It will be zero when $x(n)$ and $y(n)$ are statistically independent, and we say that the two sequences are uncorrelated. Hence, the cross-covariance function reveals the degree of similarity between two sequences. When they are partially correlated, the cross-covariance function will peak at some value of τ and tend to fall away to zero for large values of τ.

8.6 THE POWER SPECTRUM

The difficulty which random DT signals pose, and which is the focus of this chapter, is that they are most generally infinite-duration, infinite-energy sequences. Hence, the z-transform and the Fourier transform of such signals do not exist.

In the preceding sections, however, we have developed signal descriptions in terms of the autocorrelation and autocovariance functions, $R_{xx}(\tau)$ and $\gamma_{xx}(\tau)$ respectively. These embody time-domain information by virtue of the fact that they are functions of the time-shift variable τ. Further, in the case of a zero-mean random sequence, $R_{xx}(\tau)$

tends to zero for large τ while this property holds for $\gamma_{xx}(\tau)$ irrespective of the mean value. Hence, $R_{xx}(\tau)$ and $\gamma_{xx}(\tau)$ serve as useful finite-duration, time-domain descriptions of $x(n)$. Because the sequence is finite-duration, it is also finite energy and its Fourier transform exists.

To keep things simple initially, let us suppose that the random signal is a continuous-time function, $x(t)$, and its values are known over all time. That is, we are dealing with ensemble values of the random process at this stage. Then the Fourier transform of its ACF, treating the time shift τ as a continuous variable, is:

$$P_{xx}(j\omega) = \int_{-\infty}^{\infty} R_{xx}(\tau)e^{-j\omega\tau}d\tau$$

which is a function of the continuous frequency variable ω.

Since the ACF is an even function of τ, it follows that $P_{xx}(j\omega)$ is real and even. Accordingly, we can drop the j from its argument and write it $P_{xx}(\omega)$. Its Fourier transform simplifies to:

$$\begin{aligned} P_{xx}(\omega) &= \int_{-\infty}^{\infty} R_{xx}(\tau)\cos(\omega\tau)d\tau \\ &= 2\int_{0}^{\infty} R_{xx}(\tau)\cos(\omega\tau)d\tau \end{aligned}$$

The inverse Fourier transform is:

$$\begin{aligned} R_{xx}(\tau) &= \frac{1}{2\pi}\int_{-\infty}^{\infty} P_{xx}(\omega)e^{j\omega\tau}d\omega \\ &= \frac{1}{\pi}\int_{0}^{\infty} P_{xx}(\omega)\cos(\omega\tau)d\omega \end{aligned}$$

Hence:

$$\begin{aligned} R_{xx}(0) &= \frac{1}{\pi}\int_{0}^{\infty} P_{xx}(\omega)d\omega \\ &= E[x^2(t)] \qquad (\geq 0) \end{aligned}$$

and the area under the curve of $P_{xx}(\omega)$ versus ω is a direct measure of the average power of $x(t)$. Similarly, if we integrate between finite limits $\Omega \pm \delta\omega/2$ say, we obtain the average power in the frequency band $\delta\omega$ centred on Ω. Hence, letting $\delta\omega$ tend to zero,

$P_{xx}(\omega)$ describes the way that the power of the random process is distributed across frequency. Accordingly, it is called the *power density spectrum* (or power spectrum) of the process. Again, the now-familiar concept of density indicates that it is not sensible to think of the power at a particular frequency, only within a band of frequencies. Since power is non-negative, so too is the function $P_{xx}(\omega)$. Note that the units of power spectral density are V^2/Hz.

The fact that the power density spectrum and ACF are related as a Fourier transform pair:

$$\begin{aligned} P_{xx}(\omega) &= \mathcal{F}[R_{xx}(\tau)] \\ R_{xx}(\tau) &= \mathcal{F}^{-1}[P_{xx}(\omega)] \end{aligned}$$

is called the *Wiener-Khinchin theorem.* This definition in terms of the ACF, although standard in the literature, can cause difficulties when the random signal does not have zero mean. The effect of non-zero mean is to introduce a delta function into the power density spectrum at $\omega = 0$. This difficulty is avoided if we use the autocovariance function instead.

Although the Wiener-Khinchin theorem is framed in terms of random signals, it would be only reasonable to expect that it also applied to deterministic signals. This is easy to verify for the particular case of a sine wave. According to equation (8.24), the ACF has the form of a cosine wave, whose Fourier transform is a delta function. By Parseval's theorem, equation (1.20), the power spectrum of a signal consisting of just one harmonic consists of a single line, i.e. it is a delta function.

The above definition of the power density spectrum can be extended to two jointly stationary random processes, $x(t)$ and $y(t)$, having CCF $R_{xy}(\tau)$ to give:

$$P_{xy}(j\omega) = \int_{-\infty}^{\infty} R_{xy}(\tau)e^{-j\omega\tau}d\tau$$

Here, $P_{xy}(j\omega)$ is the *cross-power density spectrum.* In this case, since $R_{xy}(\tau)$ is not even, $P_{xy}(j\omega)$ is not generally real.

We have developed the above theory for continuous-time signals, $x(t)$ and $y(t)$, for simplicity. It should be clear, however, that the generalisation to discrete-time signals is straightforward. In principle, we can define the power spectrum of $x(n)$ using the DFT as:

$$P_{xx}(m) = \sum_{\tau=0}^{N-1} R_{xx}(\tau)e^{-2\pi jm\tau/M}$$

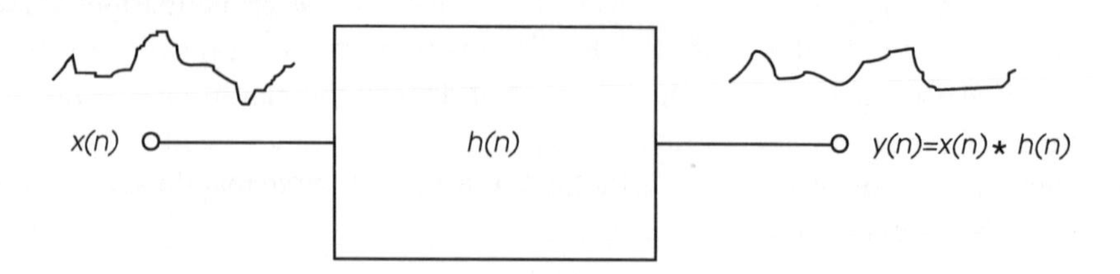

Fig. 8.8 Linear time-invariant system subjected to random input, $x(n)$.

where we revert to τ as a discrete variable. Note that here frequency is no longer continuous so that $P_{xx}(m)$ is no longer a density. Rather $P_{xx}(m)$ corresponds to the power within a frequency band of width $2\pi/MT$ hertz, the frequency spacing of the DFT.

8.7 RANDOM SIGNALS AND LINEAR SYSTEMS

One of the major applications of power density spectral functions is in system identification, whereby we determine an initially-unknown system function $H(j\omega)$. We will deal with DT signals, assuming them for the purposes of this section to be infinite in duration.

Consider a linear time-invariant system whose impulse response is $h(n)$, subjected to random input $x(n)$, as depicted in figure 8.8. The output is given by the convolution-sum as:

$$\begin{aligned} y(n) &= x(n) * h(n) \\ &= \sum_{k=-\infty}^{\infty} x(k)h(n-k) \\ &= \sum_{k=-\infty}^{\infty} x(n-k)h(k) \end{aligned} \tag{8.25}$$

so that the output is also random.

Multiplying equation (8.25) by $x(n-\tau)$ and taking expected values with respect to n:

$$E[x(n-\tau)y(n)] = \sum_{k=-\infty}^{\infty} E[x(n-\tau)x(n-k)]h(k)$$

Replace the dummy variable n by $(n+\tau)$:

$$E[x(n)y(n+\tau)] = \sum_{k=-\infty}^{\infty} E[x(n)x(n+\tau-k)]h(k)$$

By definition of the CCF and ACF:

$$R_{xy}(\tau) = \sum_{k=-\infty}^{\infty} R_{xx}(\tau-k)h(k)$$

$$\text{and, hence } R_{xy}(\tau) = R_{xx}(\tau) * h(\tau)$$

Taking Fourier transforms:

$$P_{xy}(j\omega) = P_{xx}(\omega)H(j\omega) \tag{8.26}$$

and:

$$H(j\omega) = \frac{P_{xy}(j\omega)}{P_{xx}(\omega)} \quad \text{for } P_{xx}(\omega) \neq 0$$

So provided the input has non-zero spectral density at all frequencies of interest, $H(j\omega)$ can be found in magnitude and phase from the cross-power density spectrum and the input power spectrum. Recall from above that a totally uncorrelated noise signal has an ACF which is a (scaled) delta function, $k\delta_0$, the Fourier transform of which (i.e. the input power spectrum) is identically equal to k. Hence, such an input signal allows the system to be identified very conveniently. This is actually just another statement of the fact that a linear time-invariant system is fully specified by its impulse response, since the ACF of a delta function (impulse) is also a delta function.

We have assumed here that we have some system, with system function $H(j\omega)$, whose input is accessible – so that we can apply a suitable random excitation and compute the cross-power spectrum. In many important cases, this will not be so. Suppose, for instance, we treat human speech or electrical potentials from brain activity recorded on the scalp as output signals from some system which we wish to identify. It may not be clear exactly what the input to the system is. Even if it were, applying a random input to it may be out of the question. In such cases, it may still be possible to characterise the system in some useful way, as we now show.

Suppose we multiply equation (8.25) by $y(n+\tau)$ and take expected values:

$$E[y(n+\tau)y(n)] = \sum_{k=-\infty}^{\infty} E[y(n+\tau)x(n-k)]h(k)$$

Replace the dummy variable n by $(n - \tau)$:

$$E[y(n)y(n-\tau)] = \sum_{k=-\infty}^{\infty} E[y(n)x(n-\tau-k)]h(k)$$

Again, by definition of the CCF and ACF:

$$R_{yy}(-\tau) = \sum_{k=-\infty}^{\infty} R_{yx}(-\tau-k)h(k)$$

But, since:

$$\begin{aligned} R_{yy}(-\tau) &= R_{yy}(\tau) \\ \text{and } R_{yx}(-\tau) &= R_{xy}(\tau) \end{aligned}$$

we have:

$$\begin{aligned} R_{yy}(\tau) &= \sum_{k=-\infty}^{\infty} R_{xy}(\tau+k)h(k) \\ &= R_{xy}(\tau) * h(-\tau) \end{aligned}$$

Taking Fourier transforms:

$$P_{yy}(\omega) = P_{xy}(j\omega)\tilde{H}(j\omega)$$

Substituting for the cross-power density spectrum from equation (8.26):

$$\begin{aligned} P_{yy}(\omega) &= P_{xx}(\omega)H(j\omega)\tilde{H}(j\omega) \\ &= P_{xx}(\omega)|H(j\omega)|^2 \end{aligned} \tag{8.27}$$

Clearly, $|H(j\omega)|^2$ is a system descriptor of some importance, detailing how the input power is modified as a function of frequency to produce output power. It is called the *system power spectrum*. If the input to the system is inaccessible so that $P_{xx}(\omega)$ is unknown, we can always assume it to be identically 1, corresponding to an $x(n)$ with a delta-function ACF (either an impulse or uncorrelated noise of appropriate amplitude). This amounts to modelling whatever process produces the (inaccessible) 'input' to the system as part of the system itself.

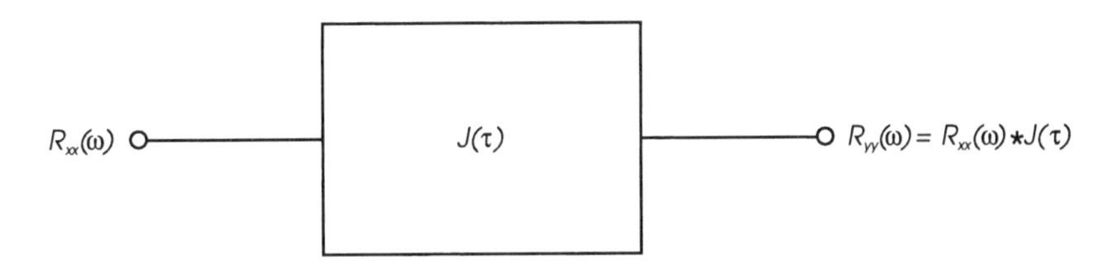

Fig. 8.9 The function $J(\tau)$ describing a linear time-invariant system can be interpreted as the system's 'impulse response' in the autocorrelation domain. That is, $J(\tau)$ is the ACF of the impulse response.

If we take the inverse Fourier transform of equation (8.27), we have:

$$\begin{aligned} R_{yy}(\tau) &= R_{xx}(\tau) * J(\tau) \\ \text{where } J(\tau) &= \mathcal{F}^{-1}\left[|H(j\omega)|^2\right] \end{aligned} \tag{8.28}$$

Hence:

$$\begin{aligned} \mathcal{F}[J(\tau)] &= H(j\omega)\tilde{H}(j\omega) \\ \text{so that } J(\tau) &= h(\tau) * \tilde{h}(\tau) \\ &= \sum_{k=-\infty}^{\infty} h(k)\tilde{h}(\tau - k) \\ &= \sum_{k=-\infty}^{\infty} h(k)h(k - \tau) \\ &= R_{hh}(-\tau) \\ &= R_{hh}(\tau) \end{aligned}$$

That is, $J(\tau)$ is the autocorrelation function of the system's impulse response:

$$J(\tau) = E[h(n)h(n + \tau)]$$

Equation (8.28) has a useful interpretation in system terms, as shown in figure 8.9.

8.8 ESTIMATION FROM FINITE-LENGTH SEQUENCES

We have shown that both the ACF and its Fourier transform (the power spectrum) are important descriptors of a random process. For instance, we have used these ideas to show how we can characterise a system with unknown properties subject to a random input. However, the ACF is defined (equation (8.20)) as a statistical average over all time:

$$R_{xx}(\tau) = E[x(n)x(n+\tau)] = \lim_{N\to\infty} \frac{1}{N} \sum_{n=-N/2}^{N/2-1} x(n)x(n+\tau)$$

In a practical situation, we have only finite-duration, length-N sequences available to us. If the process is ergodic, then one way to estimate the ACF from a single such realisation would be as:

$$\hat{R}_{xx}(\tau) = \frac{1}{N} \sum_{n=0}^{N-1} x(n)x(n+\tau) \tag{8.29}$$

The obvious questions which arise are these. How good an estimator of the ACF is equation (8.29), and how good an estimate of the power spectrum do we obtain by applying the DFT to it? The answers to these questions lie in the realms of statistical estimation theory, treatment of which is beyond the scope of this book. Our purpose here is to introduce some of the issues involved in estimating the power spectra of random signals, as a precursor to further study.

Suppose we wish to estimate some parameter α as $\hat{\alpha}$ from random sample $\{x(n)\}$, $0 \leq n < N$. It follows that $\hat{\alpha}$ is a function of $x(0), x(1), \ldots, x(N-1)$ so that it too is a random variable. Its PDF will be $p(\hat{\alpha})$. From earlier, if the mean of this PDF, $E[\hat{\alpha}]$, is equal to α, then $\hat{\alpha}$ is an unbiased estimator of α. We define the bias, B, as:

$$B = \alpha - E[\hat{\alpha}]$$

The variance of the PDF is:

$$\sigma^2(\alpha) = E[(\hat{\alpha} - E[\hat{\alpha}])^2]$$

A good estimator will have very small associated variance as well as being unbiased. If both the bias and variance tend to zero as N tends to infinity, we say that the estimator is *consistent*.

How good an estimator of the ACF is equation (8.29) above? Since $\{x(n)\}$ is of finite-length N, only $(N - |\tau|)$ values enter into the summation – the others being zero. Hence:

$$
\begin{aligned}
E[\hat{R}_{xx}(\tau)] &= \frac{1}{N}\sum_{n=0}^{N-|\tau|-1} E[x(n)x(n+\tau)] \\
&= \frac{1}{N}\sum_{n=0}^{N-|\tau|-1} R_{xx}(\tau) \\
&= \frac{N-|\tau|}{N} R_{xx}(\tau) \qquad |\tau| \le N
\end{aligned}
$$

and the estimator of equation (8.29) is biased, since its mean is not equal to the ACF. However, the estimate of the mean improves as N increases relative to $|\tau|$: hence, this is called *asymptotically unbiased*. Clearly, though, a better (unbiased) estimator would be:

$$
\hat{R}'_{xx}(\tau) = \frac{1}{N-|\tau|}\sum_{n=0}^{N-1} x(n)x(n+\tau) \tag{8.30}
$$

Note that essentially identical reasoning lies behind the definition of the sample standard deviation in equation (8.8) earlier – see also exercise 8.6.

In fact, both estimates of the ACF that we have examined are consistent. From equations (8.29) and (8.30):

$$
\hat{R}'_{xx}(\tau) = \frac{N}{N-|\tau|}\hat{R}_{xx}(\tau)
$$

Assuming zero mean for simplicity, the variance of $\hat{R}'_{xx}(\tau)$ is $[N/(N-|\tau|)]^2$ (≥ 0) times that of $\hat{R}_{xx}(\tau)$. In this respect, the latter estimate is in some sense 'better' even though it is biased. In particular, the variance of $\hat{R}'_{xx}(\tau)$ becomes very large when the shift τ approaches N and we do not obtain useful estimates of the ACF. On the other hand, the variance of $\hat{R}_{xx}(\tau)$ does not increase in this way, but the estimate of the ACF obtained tends to zero (since very few products are entered into the sum). So the bias, B, tends to the same order as the mean itself. These sort of problems can be eased by defining a circular form of autocorrelation, much as we defined a circular convolution in chapter 5 for finite-length sequences.

Although both estimates are consistent, and either unbiased or asymptotically unbiased, it does not follow that their Fourier transforms will yield good estimates of the power density spectrum. In fact, they do not. The basis of the problem is that the variance of the power spectral estimates obtained in this way does not tend to 0 as N tends to infinity, i.e. they are not themselves consistent. The answer lies in somehow smoothing the power-spectral estimates, by ensemble averaging or using windowing techniques. This is a large and complex topic and will not be dealt with further here.

8.9 SUMMARY

In this, the last chapter of the book, we have drawn a distinction between the sort of deterministic signals which we had concentrated on previously and the important class of random signals. A deterministic signal is one for which future values can be correctly predicted from a mathematical model of the signal's generation process, with observed, past values used to find the parameters of the model. By contrast, a random – or stochastic – signal is unpredictable because its generation process is too complex or poorly understood to be modelled in this way. Clearly, the boundary between these two classes is somewhat problematic. It could be reasonably argued that present ignorance of the generation process of a particular signal does not preclude us from ever understanding it, and so modelling it predictively. Accordingly, the whole concept of 'randomness' is open to philosophical question. Fortunately, this is not a major concern. The concept is of proven utility and, in any event, there is little or nothing to stop us treating deterministic signals within the framework we have constructed for describing random signals. The penalty for so doing is that the description is considerably less precise.

Random signals are important in the study of discrete-time signals and systems from many points of view. Often the signals which we encounter in nature, and which we wish to understand, either have a random component or are stochastic in nature to a high degree. For instance, electronic processing of an otherwise deterministic signal can add thermal noise while (from the perspective of current knowledge at least) the electrical potentials recorded on the human scalp as a result of brain activity appear almost entirely random.

What then are the particular difficulties posed by random signals? Throughout this book, we have used the z-transform and related Fourier descriptions to characterise discrete-time signals (and systems). In order for the z-transform and Fourier transform to exist, certain convergence criteria must be met. These amount to requiring that the signals are either of finite duration with finite energy, or are periodic with finite power. Random signals, however, are most naturally modelled as aperiodic, infinite in duration, and having infinite energy (but finite power).

The key to the treatment of random signals is to describe them in terms of their statistical averages – or expected values – over a long time. Immediately we confront a problem since the signal is infinite in duration yet, in a practical situation, we cannot deal with such a sequence. We distinguish, therefore, between the entire signal – the ensemble – and sample sequences, or realisations, drawn from the ensemble. Two especially useful average descriptors of the ensemble are the mean and variance, corresponding to the first and second moments respectively. These must be estimated from finite-duration sample sequences, either by averaging over time or by aligning multiple realisations and ensemble averaging. A random process for which the result of ensemble averaging is independent of the particular time index at which the average is taken is called stationary. If ensemble-average values are not different from time-average values – apart from variation arising from statistical sampling effects – the process is called ergodic. We have assumed ergodicity throughout. The importance of this property is that we need only a single realisation, albeit of appropriately long duration, to describe the random signal.

While the mean and variance, and possibly higher-order moments too, portray useful information about specific statistical attributes of a random signal, the probability density function (PDF) embodies information about all the moments. Since the amplitude values, $x(n)$, of the random sequence $\{x(n)\}$ are assumed to be continuous, the notion of the probability of $x(n)$ being exactly equal to a specific value is not useful. Since there is an infinity of possible specific values that $x(n)$ could attain, such a probability is always equal to zero. Rather, we think in terms of the probability of $x(n)$ being within a range δy about some value, y. As the width of this range reduces to zero, so does this latter probability by the same token as previously. However, the ratio of this probability to the reducing width δy tends to a finite value called the probability density. The way that this density varies as a function of y is the PDF. The PDF is not unique to a particular random process: rather it describes a class of processes.

We then studied two such classes of random process – described by a rectangular PDF and a Gaussian (normal) PDF respectively. Sample sequences generated by these processes were presented, and PDFs estimated from these realisations. These were shown to be reasonable approximations. To gain additional insight into the use and properties of the PDF, we also determined this function for a familiar deterministic signal – the sine wave. This emphasised that although information about the mean and average power were conserved, time-domain properties such as the phase were entirely absent.

We next studied the autocorrelation function (ACF), a signal description which attempts to conserve information about time-domain properties. The ACF, $R_{xx}(\tau)$, is the statistical expectation of the product of $\{x(n)\}$ and a version of $\{x(n)\}$ time-shifted by τ sample periods. Assuming $x(n)$ values to be real, it is a real and even function. The ACF attains its maximum value at $\tau = 0$: this value is an extremum and is equal to the average power in the signal. If the value of the ACF at some $\tau \neq 0$ is zero, then values of $x(n)$ separated by this τ are said to be uncorrelated. This is a related concept to statistical independence: we ignored any distinction and took the two to be synonymous. Since a truly random signal is uncorrelated for all values of τ except $\tau = 0$, the ACF of such a signal is a Kronecker delta function (whose height equals the average power). More generally, the amplitude of the ACF for partially-correlated random signals will tend to fall off as τ increases, since any correlation that does exist is usually short-range. Hence, we have the important property that the ACF can be modelled as a finite-duration sequence whose Fourier transform accordingly exists.

Subsequently, we studied the correlation of $\{x(n)\}$ not with itself but with some other random signal $\{y(n)\}$. This led us to define a cross-correlation function (CCF), denoted $R_{xy}(\tau)$, which reduces to the ACF when $\{y(n)\}$ is equal to $\{x(n)\}$. Unlike the ACF, the CCF is not an even function of τ, but $R_{xy}(\tau) = R_{yx}(-\tau)$. Again, if $\{x(n)\}$ and $\{y(n)\}$ are uncorrelated then $R_{xy}(\tau) = 0$ while for partially-correlated sequences $R_{xy}(\tau)$ usually falls off with τ. Hence, the CCF can also be modelled as a finite-duration sequence.

If the ACF, $R_{xx}(\tau)$, does indeed have finite duration, its Fourier transform exists and is called the power density spectrum, $P_{xx}(\omega)$, or power spectrum. It describes how the signal power is distributed across frequency, and the integral of $P_{xx}(\omega)$ over all frequency is a measure of the (total) average power. Since frequency is a continuous variable, the power at any particular frequency is necessarily zero. Hence, we make use of the same concept of density as in the PDF. The power spectrum is real, non-negative and even: it is an indirect description of a random process – the Fourier transform of the ACF. Strictly, it *cannot* be found from the FT of the random signal itself (using Parseval's theorem, as we would do for a deterministic signal) since the random signal does not, in general, have an FT. Similarly, we can take the Fourier transform of the CCF to give the cross-power density spectrum, $P_{xy}(j\omega)$. Since the CCF is not an even function of τ, this is a real function of frequency. The power spectrum and cross-power spectrum are very useful in system identification, enabling us to determine the system function $H(j\omega)$ for a flat-spectrum input from the input-output cross-power density spectrum.

Finally, we returned to the practical issue of estimating these important descriptors of a random process – the ACF, CCF, power spectrum etc. – from finite-length sample sequences. These estimations are themselves random and have a PDF, whose mean should be equal to the parameter being estimated and whose variance should be small. The concepts of unbiased and consistent estimators were outlined, illustrated by the autocorrelation function. Although it is possible to obtain unbiased and consistent estimators of the ACF, it is an unfortunate fact that their Fourier transforms do not yield good (consistent) estimates of the power spectrum. The answer to this problem lies in smoothing the estimates appropriately.

8.10 Exercises

8.1 Show that the variance of a random process $x(n)$ is given as:

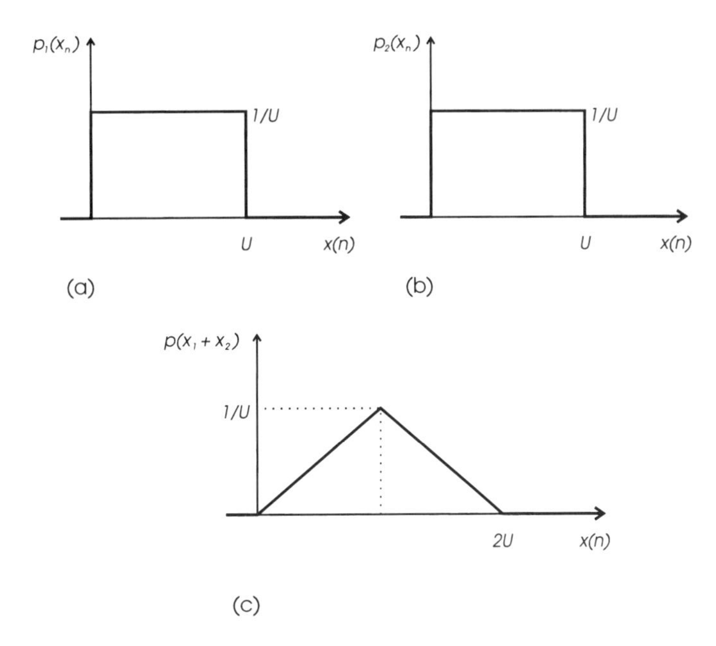

Fig. 8.10 (a) Rectangular PDF of random process $x_1(n)$, (b) identical rectangular PDF of random process $x_2(n)$, and (c) PDF of $y(n) = x_1(n) + x_2(n)$.

$$\sigma^2 = E(x_n^2) - \mu^2$$

where $E(x_n^2)$ is the expected value of $x^2(n)$.

8.2 (a) Two independent random processes $x_1(n)$ and $x_2(n)$ have PDFs, $p_1(x_n)$ and $p_2(x_n)$. Show that the PDF of their sum, $y(n) = x_1(n) + x_2(n)$, is given by the convolution:

$$p(y_n) = \int_{-\infty}^{\infty} p_1(z) p_2(y_n - z) dz$$

(b) If the PDFs of $x_1(n)$ and $x_2(n)$ are identical, and rectangular in the range $(0,U)$ as depicted in figure 8.10(a) and (b), show that the PDF of $y(n)$ is as depicted in figure 8.10(c). Use either a geometrical construction or the convolution integral. What are the mean and variance of the random process $y(n)$?

(c) By taking the convolution of the PDF in figure 8.10(c) with itself, find the PDF of the random process formed by summing four independent random signals all having

PDFs rectangular between 0 and U.
(NOTE: this exercise illustrates the important *central limit theorem* which states the probability density function of a sum of n independent random variables tends to a normal distribution as n tends to infinity. If the random variables are rectangularly distributed between 0 and 1 as here, then the mean of the normal distribution tends to $n/2$ and the variance to $n^2/12$ as n tends to infinity. A reasonable approximation to a normal distribution can be obtained with quite small values of n. In fact, the Gaussian random signal depicted in figure 8.7 was generated in this way using $n = 12$.)

8.3 (a) From the PDF for a sine wave of amplitude A, equation (8.18), confirm that the mean value of this signal is zero and its variance is equal to the long-term average power $A^2/2$. Use the standard integral formulae:

$$\int \frac{x}{\sqrt{a^2 - x^2}} dx = -\sqrt{a^2 - x^2}$$

$$\int \frac{x^2}{\sqrt{a^2 - x^2}} dx = -\frac{x}{2}\sqrt{a^2 - x^2} + \frac{a^2}{2} \sin^{-1} \frac{x}{|a|}$$

(b) Explain why, according to equation (8.18) and figure 8.5, the PDF for this signal tends to infinity at $x(n) = \pm A$ while at other values in the range $(-A, A)$ it takes a finite value.

8.4 Sketch the PDF of a sinusoidal signal contaminated by additive, zero-mean Gaussian noise. State how you would estimate (i) the amplitude of the sine wave, (ii) the variance of the noise, (iii) the signal-to-noise ratio. Also sketch the autocorrelation function and again state how you would estimate these values. Comment on the relative merits of the two approaches in practice.

8.5 (a) Show that the cross-correlation and cross-covariance functions of $x(n)$ and $y(n)$ satisfy:

$$R_{xy}(\tau) = R_{yx}(-\tau)$$

$$\gamma_{xy}(\tau) = \gamma_{yx}(-\tau)$$

(b) Show that the cross-correlation function satisfies the inequalities:

$$|R_{xy}(\tau)|^2 \leq R_{xx}(0)R_{yy}(0)$$

$$|R_{xy}(\tau)| \leq \frac{1}{2}\left[R_{xx}(0) + R_{yy}(0)\right]$$

and derive corresponding inequalities for the cross-covariance function.

(c) Show that the cross-power density spectrum satisfies:

$$P_{xy}(j\omega) = \tilde{P}_{yx}(-j\omega)$$

8.6 Show that equation (8.8) is an unbiased estimator of the standard deviation.

Index